ISBN 978-0-265-26823-0
PIBN 10283998

This book is a reproduction of an important historical work. Forgotten Books uses state-of-the-art technology to digitally reconstruct the work, preserving the original format whilst repairing imperfections present in the aged copy. In rare cases, an imperfection in the original, such as a blemish or missing page, may be replicated in our edition. We do, however, repair the vast majority of imperfections successfully; any imperfections that remain are intentionally left to preserve the state of such historical works.

FB &c Ltd, Dalton House, 60 Windsor Avenue, London, SW19 2RR.
Company number 08720141. Registered in England and Wales.

For support please visit www.forgottenbooks.com

AN ACCOUNT OF THE

ALCYONARIANS

COLLECTED BY THE

ROYAL INDIAN MARINE SURVEY SHIP

INVESTIGATOR

IN THE

INDIAN OCEAN

BY

J. ARTHUR THOMSON, M.A.,

REGIUS PROFESSOR OF NATURAL HISTORY IN THE UNIVERSITY OF ABERDEEN

AND

W. D. HENDERSON, M.A., B.Sc.,

CARNEGIE FELLOW, UNIVERSITY OF ABERDEEN

I. THE ALCYONARIANS OF THE DEEP SEA

CALCUTTA:

PRINTED BY ORDER OF THE TRUSTEES OF THE INDIAN MUSEUM

1906

INTRODUCTION.

THIS memoir contains a description of the rich collection of Deep-Sea Alcyonarians made by the Royal Indian Marine Survey Ship "Investigator" in the Indian Ocean.[1] It will be followed by another dealing with the littoral forms.

The collection includes eighty-six species, of which sixty-one (and three varieties) are new. It has been found necessary to establish five new genera,—*Stereacanthia* and *Agaricoides* in the family Nephthyidæ, subfamily Siphonogorginæ; *Acanthomuricea* and *Calicogorgia* in the family Muriceidæ; and *Thesioides* in the family Kophobelemnonidæ. The general position of the new forms is as follows:

	Total Number of Genera.	New Genera.	Total Number of Species.	New Species.
Order I. Stolonifera	1 (*Sympodium*)	0	6	6
„ II. Alcyonacea	6	2	9	8
„ III. Pseudaxonia	5	0	6	3
„ IV. Axifera	22	2	35	22 and 2 varieties
„ V. Stelechotokea	15	1	30	22 and 1 variety
Totals . .	49	5	86	61 and 3 varieties

[1] I wish to thank Professor A. Alcock, LL.D., F.R.S., Superintendent of the Indian Museum, for giving me the opportunity of studying this fine collection of beautiful forms, and the Trustees of the Indian Museum for a grant towards expenses. Through the Carnegie Trust I was able to secure the help of Mr. W. D. Henderson, joint author of this memoir, whose work in this connection was in part done during his tenure of a Carnegie Scholarship and Fellowship. We are greatly indebted to Mr. George Davidson, artist, for the skill and patience which he has bestowed on the illustrations. We think that the engraver, Mr. Edwin Wilson, also deserves to be congratulated on his success, for the coloured plates in particular presented unusual difficulties which have been happily overcome. Finally, I wish to acknowledge my great indebtedness to my private assistant, Mr. James J. Simpson, M.A., B.Sc., without whose aid the completion of the memoir would have been much delayed. His skill will be seen in the description of the remarkable genus *Agaricoides*. Dr. J. Versluys, of Amsterdam, most generously gave me his notes on *Caligorgia indica*, and Professor S. J. Hickson, F.R.S., of Victoria University, Manchester, was kind enough to allow me to consult his fine collection, and to lend me some of the literature.—J. A. T.

New Types.

The genus *Stereacanthia*, from the Andamans, is a Siphonogorgid in the vicinity of *Lemnalia.* A bare, densely spiculose trunk, made up of large longitudinal canals, with thin spiculose walls, bears a branched polyparium with the polyps disposed singly or in small crowded bundles; the aboral bands of spicules on the infolded tentacles form a simple pseudo-operculum; the spicules are warty spindles or golf-club forms, and there are no quadriradiate double-stars as in *Lemnalia.*

The genus *Agaricoides*, from 6° 31′ N., 79° 33′ 45″ E., is a remarkable Siphonogorgid perhaps distinctly related to *Lemnalia* (Gray, emend. Bourne), but quite unlike any other type known to us. It is unbranched, mushroom-like, with complex octagonal verrucæ, pedicelled anthocodiæ, introversible zooids, a tentacular operculum, echinate spindles and hockey-club forms, and many peculiarities of structure.

The genus *Acanthomuricea*, represented by *A. ramosa* from 7° 55′ N., 81° 47′ E., 506 fathoms, and *A. spicata* from 6° 31′ N., 79° 38′ 45″ E., 401 fathoms, is a Muriceid, perhaps related to *Placogorgia* (Wright and Studer). The two species are upright colonies, irregularly branched in one plane, with thin bark-like cœnenchyma of rough imbricating scales, with prominent verrucæ on all sides, with conical tentacular opercula, and with very heterogeneous spiculation.

The genus *Calicogorgia*, represented by *C. investigatoris* from 11° 14′ 30″ N., 74° 57′ 15″ E., 68–148 fathoms, and *C. rubrotincta* from the Bay of Bengal, 88 fathoms, is a Muriceid, probably related to Verrill's somewhat vaguely defined *Anthogorgia.* The colonies are irregularly branched in one plane, the verrucæ are prominent with spicules in eight bands, with a conical operculum consisting of a crown and points, with warty spindles straight or curved.

The genus *Thesioides*, from 18° 0′ 15″ N., 93° 30′ 45″ E., 448 fathoms, and 16° 25′ N., 93° 43′ 30″ E., 463 fathoms, is a Kophobelemnonid, near *Bathyptilum*, with a greatly elongated slender rachis borne by a short stalk without pinnules, with long slender autozooids without calyces and without any spicules.

Notes on some new Species.

Sympodium, sp. We have described six new species of *Sympodium*, but it seems that in this genus, as in other simple forms like *Clavularia*, there is considerable variability in the specific characters. It may also be that a colony differs considerably according to the substratum on which it grows,—a vegetable axis in *S. indicum* and *S. decipiens*, an Antipatharian axis in *S. granulosum*, a sponge spicule in *S. incrustans*, a cluster of sponge spicules in *S. tenue*, a sponge skeleton in *S. pulchrum.* It is not easy at present to give distinctive diagnoses of our six forms, and yet the *tout ensemble* of the characters of each results in a

'Clavulaires des Posidonies' are oviparous. *Sympodium* (*Alcyonium*) *coralloides* is, according to these authors, viviparous. Koren and Danielssen state that three species of *Nephthya*, found at depths of 269–761 fathoms, are viviparous. These are, I believe, the only authenticated cases of viviparity among Alcyonarians hitherto recorded. *Gorgonia capensis* affords the first instance of viviparity that I have come across in my studies of Alcyonarians."

In Professor W. A. Herdman's collection from Ceylon we also found embryos *in situ* in *Gorgonia capensis*, Hickson. Corroborating Marion and Kowalewsky, we found embryos in *Clavularia pregnans*, Th. and H., and *Clavularia parvula*, Th. and H., collected by Mr. Cyril Crossland, B.Sc., from Zanzibar and Cape Verde Islands respectively.

In the present collection we have found embryos—blastulæ, gastrulæ, and slightly more advanced stages in the following species:

Sarcophytum aberrans, n. sp. (From 254 fathoms.)
Chrysogorgia flexilis, Wright and Studer. (From 401, 457–589, 606, and 669 fathoms.)
Ceratoisis gracilis, n. sp. (From 270–45 fathoms.)
Paramuricea indica, n. sp. (From 265 fathoms.)
Distichoptilum gracile, Verrill. (From 836 fathoms.)
Umbellula elongata, n. sp. (From 360 fathoms.)
Funiculina gracilis, n. sp. (From 406 fathoms.)
Pennatula indica, n. sp. (From 463, 487, and 824 fathoms.)

Meanwhile Mr. James J. Simpson, M.A., has also found embryos in specimens of *Isis hippuris*, included in the littoral collection. (See *Journ. Linn. Soc.* (Zool.) 1906.)

It is therefore clear that viviparity is by no means uncommon in Alcyonarians, and it will be interesting to discover if it is particularly characteristic of deep-sea forms. (See Thomson and Henderson, *Zool. Anzeiger*, 1906.) We hope, when time permits, to study the embryos more carefully.

Geographical Distribution.

There are 25 species in the collection which have been previously described, and we give a list of the localities where these have been found, reserving further discussion for the second memoir on the littoral forms. It may be noted that of the 25—

3	were included	in the	"Challenger"	collection.
6	,,	,,	Siboga	,,
4		,,	Herdman's Ceylon	,,
3		,,	Gardiner's Maldives	,,
3	,,	,,	Willey's New Britain	,,
1		,,	Crossland's Zanzibar	,,

1 was included in the "Alert" collection.
3 were ,, ,, Funafuti ,,
1 ,, ,, South African ,,
1 ,, ,, "Scotia" ,,
2 ,, ,, "Hirondelle" ,,

Chironephthya variabilis, Hickson. Ceylon (Herdman), Maldives (Gardiner).

Suberogorgia köllikeri, Wright and Studer. Japan (Challenger), Ceylon (Herdman), Zanzibar (Crossland).

Keroëides koreni, Wright and Studer. Japan (Challenger), Funafuti (Hiles).

Keroëides gracilis, Whitelegge. Funafuti (Whitelegge), British New Guinea (Willey), Ceylon (Herdman).

Lepidogorgia verrilli, Wright and Studer. Japan (Challenger), Macassar Straits (Siboga).

Chrysogorgia orientalis, Versluys. East Indian Archipelago (Siboga).

Chrysogorgia flexilis, Wright and Studer. Chiloe (Challenger), East Indian Archipelago (Siboga).

Acanella rigida, Wright and Studer. Off Banda (Challenger).

Stachyodes allmani, Wright and Studer. Fiji (Challenger), Celebes (Siboga).

Thouarella moseleyi, Wright and Studer. Kermadec Islands (Challenger), Flores (Siboga).

Caligorgia flabellum, Ehrenberg. Near Mauritius, East Pacific off Central America, Japan, East Indian Archipelago (Siboga).

Acanthogorgia aspera, Pourtales. Havana, Azores (Hirondelle); as = *A. spinosa*, Hiles ?, Sandal Bay, Lifu (Willey).

Acamptogorgia bebrycoides, von Koch. Mediterranean (von Koch), Azores (Hirondelle).

Callistephanus koreni, Wright and Studer. Off Ascension (Challenger).

Nicella flabellata (Whitelegge) = *Verrucella flabellata*, Whitelegge. Funafut (Whitelegge).

Juncella elongata, Pallas. Atlantic (Pallas), West Indies (Ellis and Solander); variety from N.E. coast of Australia (Ridley); variety, *capensis*, Algoa Bay (Hickson).

Scirpearella moniliforme, Wright and Studer. Amboina (Challenger).

Telesto arthuri, Hickson and Hiles. Blanche Bay, New Britain (Willey).

Telesto rubra, Hickson. Maldives (Gardiner), Ceylon (Herdman).

Distichoptilum gracile, Verrill. Nantucket (Verrill), 0° 4′ S., 90° 24′ 30″ W., 61° 39′ N., 17° 10′ W. (Jungersen); 23° 59′ N., 108° 40′ W., 1° 7′ N., 80° 21′ W. (Studer).

Kophobelemnon burgeri, Herklots. Japan.

Umbellula durissima, Kölliker. Japan, S. of Yeddo (Challenger), 48° 06′ S., 10° 5′ W. (Scotia).

Anthoptilum murrayi, Kölliker. N. Atlantic, S. of Halifax (Challenger); E. coast of N. America (Verrill); Bay of Gascony, S. of Iceland (Jungersen).

Funiculina quadrangularis (Pallas) = *Leptoptilum gracile*, Kölliker. New Zealand, as *L. gracile* (Challenger); as *F. quadrangularis*, North Sea, Atlantic Ocean, European and American sides, Mediterranean, etc.

Pavonaria willemoësii (Kölliker) = *Microptilum willemoësii*, Kölliker. As *M. willemoësii* from south of Yeddo (Challenger).

Among the new facts of distribution, perhaps the following are of most interest:

Stachyodes allmani, Wright and Studer. From the Laccadive Sea (Investigator); previously from the reefs, Fiji.

Callistephanus koreni, Wright and Studer. From the Andaman Sea (Investigator); previously from off Ascension.

Juncella elongata, Pallas. From the Bay of Bengal (Investigator); previously from Atlantic, West Indies, N.E. coast of Australia, and Algoa Bay.

Distichoptilum gracile, Verrill. From Investigator Station 231, 7° 34′ 30″ N., 76° 08′ 23″ E., and 321, 5° 4′ 8½ N., 80° 22′ E.; previously from North Atlantic, S.W. of Nantucket Island, etc.

Umbellula durissima, Kölliker. From Laccadives (Investigator); previously from S. of Yeddo and Antarctic.

Anthoptilum murrayi, Kölliker. From Investigator Station 104, 11° 12′ 47″ N., 74° 25′ 30″ E.; previously from N. Atlantic, Bay of Gascony, S. of Iceland.

Funiculina[1] *quadrangularis* (Pallas) = *Leptoptilum gracile*, Kölliker. From Bay of Bengal ("Investigator"), as *Leptoptilum gracile*, and previously from New Zealand. *F. quadrangularis*, previously from North Sea, Atlantic, Mediterranean, etc.

Pavonaria[1] *willemoësii* (Kölliker) = *Microptilum willemoësii*, Kölliker. From Andaman Sea; previously as *M. willemoësii* from Japan.

The wide distribution of some deep-sea types is thus well illustrated.

Some Matters of Detail.

It may be convenient to direct attention here to some matters of detail that are of general interest.

The siliceous axis which forms the support of *Sarcophytum aberrans*, n. sp., is 300 mm. in length by 2–3 mm. in breadth, and is probably the huge spicule of *Monorhaphis* or some allied sponge. (See Plate I. fig. 2c.)

Analogous, on a smaller scale, is the siliceous sponge-spicule, which serves as a support for *Sympodium incrustans*, n. sp. (See Plate II. fig. 7.)

[1] It is possible that our *Funiculina quadrangularis* is the young form of some other species of *Funiculina*, and that our *Pavonaria willemoësii* is the young form of some already known species of *Pavonaria*.

The spicules of *Chironephthya macrospiculata*, n. sp., are of unusually large dimensions, some attaining a length of 8·3 mm. (See Plate IV. fig. 10.)

In *Spongodes uliginosa*, n. sp., there are almost equally huge spicules, some 8 mm. in length.

Noteworthy is the great heterogeneity of the spicules in some of the forms, *e.g.* plates, discs, triangles, rods, spindles, and "golf-clubs" in *Acanthomuricea spicata*, n. sp., and similarly in *A. ramosa*, n. sp.

Besides the very peculiar habit,—incrusting a huge siliceous rod,—there are many interesting features in *Sarcophytum aberrans*, n. sp.: the occurrence of several sizes of autozooids, the inturning of almost the whole of a large tentacle into the stomodæum, the presence of ova and embryos in the siphonozooid canals.

In *Sarcophytum agaricoides* also there are ova in the siphonozooid cavities.

The dimorphism which Gray recorded in his ***Paragorgia nodosa*** is confirmed in *P. splendens*, n. sp. It is unique in Pseudaxonia.

In *Distichoptilum gracile*, Verrill, we have observed that there may be two *or* three siphonozooids in close connection with the autozooids.

The complex differentiation of the polyps in *Agaricoides alcocki*, Simpson, is quite unique.

Very remarkable tentacles occur in *Thesioides inermis*, n. g. et sp. (See Plate VI. figs. 1 and 2.) Those of *Protocaulon indicum*, n. sp., are also unusual (See Plate VII. fig. 3.)

The base of *Anthoptilum decipiens*, n. sp., is very characteristic in its shape, and it may be further noted that there is no evidence of an area of attachment.

In *Pteroëides triradiata* the small number (3) of supporting rays is noteworthy. It is possible that the specimen, which is only 107 mm. in length, is still young; and attention may be directed to the range of variation in the number of rays in *P. griseum*.

In regard to a collection which is a very feast of colour, we may call special attention to the exquisite colour schemes of *Pennatula veneris*, *P. pendula*, and *P. splendens*, and also to the very rich crimson-lake tint of the rachis and pinnules in *P. indica*.

The presence of numerous Foraminifera in the stomodæum of *Agaricoides alcocki* is a fact of interest.

Some of the epizoic animals are interesting, *e.g.* the peculiar Solenogaster (*Rhopalomenia gorgonophila*?) on *Acamptogorgia circium*, n. sp.; *Palythoa* and sponge on *Parisis indica*, n. sp. Between the vegetable axis and the stolons of *Sympodium indicum*, n. sp., Polychæt worms have formed burrows, and some tube-forming Polychæts are attached to the surface.

LIST OF SPECIES.

ORDER I. STOLONIFERA, Hickson.

ORDER II. ALCYONACEA, Verrill (*pro parte*).

ORDER III. PSEUDAXONIA, G. von Koch.

ORDER IV. AXIFERA, G. von Koch.

Order V. STELECHOTOKEA, Bourne.

Section ASIPHONACEA.

Section PENNATULACEA.

DESCRIPTION OF SPECIES.

ORDER I. STOLONIFERA, Hickson.

Family *CORNULARIIDÆ.*

This family is represented in the collection by six species of *Sympodium*, all of which seem to be new:

Sympodium indicum, n. sp.
,, *decipiens*, n. sp.
,, *incrustans*, n. sp.
,, *granulosum*, n. sp.
,, *tenue*, n. sp.
,, *pulchrum*, n. sp.

Sympodium indicum, n. sp. Plate II. fig. 6; Plate IX. fig. 18.

This species is represented by one greyish-white specimen which forms a complete tube round the broken fragment of a hollow vegetable axis.

The surface presents a granular appearance, and is thickly covered with large polyps which arise singly, and do not appear to be arranged in any particular manner.

The polyps consist of a retractile anthocodia and of a non-retractile calyx, the latter marked by eight longitudinal ridges which end slightly below the top. The tentacles have on their aboral surface a spiculated band which projects downwards for a short distance from their base. The calyces are 2–3·5 mm. in length; the whole polyp is about 7 mm. in length; but this does not represent the maximum, as no polyp is fully expanded. Ova are present in abundance at the base of the polyps.

The spicules are of two types, (1) prominently rough warty spindles, either straight or curved; and (2) quadriradiate forms, few in number and marked by an X-shaped marking at the origin of the rays. The following measurements were taken of length by breadth in millimetres:

1. 0·5×0·08; 0·4×0·065; 0·25×0·04.
2. 0·3×0·2; 0·3×0·13; 0·2×0·15.

Between the vegetable axis and the stolon, polychæte worms have formed burrows, and some tube-forming worms are attached to the surface.

Locality: Andamans; 265 fathoms.

Sympodium decipiens, n. sp. Plate IX. fig. 8.

This species is represented by a large greyish-white specimen creeping over and encircling a vegetable axis.

The polyps are large, occurring either singly or in groups of two or three on a membranous stolon. They consist of a retractile and a non-retractile portion, the latter with a maximum length of 7 mm.

The whole surface of the stolon and of the polyps is closely covered by large spicules which are visible to the naked eye.

The spicules are long curved spindles or rods often abruptly truncated at one end and covered with numerous rough wart-like projections. There are two types:

1. Rather thick rods, curved or straight, often bluntly rounded at one end and tapering to the other, thus tending towards a club-like form, thickly covered with rough warts and varying in length from 0·3–1·2 mm., and in breadth from 0·09–0·2 mm.

2. Longer and more slender rods or spindles with fewer and simpler projections, varying in length from 0·25–0·8 mm., and in breadth from 0·06–0·1 mm.

In both cases there is often a slight bifurcation of the ends of the spicule. This species closely resembles *Sympodium indicum*, but differs from it (1) in the larger size of the spicules, and (2) in the arrangement of the polyps.

Locality: Andamans; 271 fathoms.

Sympodium incrustans, n. sp. Plate II. fig. 7.

This species is represented by two greyish-white fragments which cover a siliceous sponge spicule.

The stolon is a thin membrane spreading over and completely surrounding the spicule. It is granular in appearance and bears few polyps, which occur irregularly at wide intervals. The calyces are marked by eight longitudinal ridges which extend up their whole length, and thus produce a crenate appearance at the margin of the cup. On the ridges the spicules are arranged with their long axes parallel, or slightly inclined, to the length of the ridge. The calyces are from 1–3·9 mm. in length and from 1–2 mm. in basal diameter.

The non-retractile portion is also closely covered with spicules which are not arranged in any particular order. Just below the tentacles they form more or less regular bands which extend up the aboral surface of each tentacle, on which the spicules are at first arranged in chevron, but are soon disposed with their long axes parallel to the length of the tentacle.

The spicules are spindles, either straight or curved, covered by thick prominent warts. The following measurements were taken of length and breadth in millimetres:

0·19×0·05; 0·23×0·05; 0·25×0·06; 0·35×0·04.

Locality: Andamans; 238–290 fathoms.

Sympodium granulosum, n. sp.

This species is represented by a single colony, with a thin membranous stolon, granular in appearance and completely surrounding an antipatharian axis.

The polyps are grouped in bundles of from 2–9 at irregular intervals, each bundle having a rough irregular ball-shaped appearance. The calyces are low and squat, rising very little above the level of the general cœnenchyma, and are covered by a rough coating of spicules.

The anthocodiæ are capable of complete retraction. The tentacles are short, and have a band of spicules on the aboral surface, consisting of two rows diverging from the middle line. The general anthocodial spicules are arranged *en chevron*. The spicules are spindles or rods, a few flat ovals, and a few quadriradiate forms. The spindles and rods may be straight or curved, and all have rough warty protuberances. The spicular measurements, length by breadth in millimetres, are as follows:

1. Spindles, 0·8×0·09; 0·5×0·08; 0·15×0·04; 0·08×0·03.
2. Rods, 0·4×0·07; 0·35×0·09; 0·3×0·1.
3. Oval forms, 0·5×0·22; 0·37×0·2; 0·3×0·2.
4. Quadriradiate and X-marked forms, 0·8×0·25; 0·3×0·2; 0·14×0·07.

Locality: Station 173; 8° 35′ 45″ N., 81° 17′ 45″ E.; 609 fathoms.

Sympodium tenue, n. sp. Plate IV. fig. 6.

To this species are referred two small colonies or fragments which spread over and enclose a number of siliceous sponge spicules.

The stolon is thin and membranous, granular in appearance, and completely surrounds several of the sponge spicules.

The polyps are scattered over the surface, occurring singly, or in twos or threes in close proximity. They consist of a non-retractile calyx and of a retractile portion, the anthocodia. When the anthocodiæ are completely retracted, the mouth of the calyx shows a number of blunt lobes or teeth.

The tentacles are of medium length, and have a band of spicules running up the aboral surface.

The calyces are from 1·5–3·9 mm. in length, and are marked by eight ridges running longitudinally, which are only faintly visible in the more contracted polyps.

The spicules of the calyx are arranged with the long axes parallel to the length of the polyp. On the anthocodiæ the spicules are arranged in a band consisting of eight inverted Vs, the points of which lie at the bases of the tentacles and project for a short distance up the aboral surface. Higher up the spicules assume a position parallel to the length of the tentacle.

The spicules are spindles, either straight or curved, with rough warty protuberances, and a very few quadriradiate forms. The following measurements were taken of length and breadth in millimetres :

Spindles, 0·1 × 0·05 ; 0·25 × 0·05 ; 0·5 × 0·07 ; 0·6 × 0·08 ; 0·65 × 0·05.

Quadriradiate forms, 0·14 from tip to tip one way.

" " 0·13 " " the other way.

Locality : Station 222 ; 13° 27′ N., 93° 14′ 30″ E. ; 405 fathoms.

Sympodium pulchrum, n. sp. Plate VI. fig. 7.

This species is represented by a large colony spreading over the surface of a sponge skeleton.

The stolon is thin and ribbon-like in parts, but it also spreads out into flat expansions, with a stringy appearance on the surface.

The polyps occur all over the surface, arising singly or in twos and threes in close proximity at the ends of the branches of the sponge on which they spread.

The calyces are usually marked by eight longitudinal ridges, which are more distinct in the larger polyps. The spicules on the calyces are arranged with the long axes parallel to the length of the calyx, and form a complete compact covering. The calyces vary in length from 1·2–7 mm., and in basal diameter from 1·1–2·5 mm.

The anthocodiæ are capable of complete retraction, and are all more or less retracted, but several even in this condition measure slightly over 3 mm., thus giving the larger polyps a total height of over a centimetre. The spicules, which form a dense covering, are arranged distally in a transverse band, from which points project at the bases of the tentacles. On the tentacles the spicules are at first arranged *en chevron*, but soon run parallel to the long axis. The band continues up the aboral surface of the tentacle and gives off obliquely transverse spicules to the sides.

The spicules are spindles, straight or curved, with rough warty protuberances. The following measurements were taken of length and breadth in millimetres :

0·8 × 0·075 ; 0·5 × 0·04 ; 0·15 × 0·04.

Locality : Station 284 ; 7°·55′ N., 81° 47′ E. ; 506 fathoms.

ORDER II. ALCYONACEA.

Family *ALCYONIDÆ.*

This family is represented by two new species of *Sarcophytum* :

Sarcophytum aberrans, n. sp.

" *agaricoides*, n. sp.

Sarcophytum aberrans, n. sp. Plate I. figs. 1, 2a, 2b, and 2c; Plate IX. figs. 7 and 11.

This species is based on two specimens which are very different from one another in habit and general appearance. The first—which diverges less than the other from the typical agariciform shape—is a coral red colony, attached by means of a thin membranous base to a piece of coral.

The stalk is 18 mm. in length, and is very much flattened owing to the collapse of the thin walls between the canals. The capitulum is small and slightly mushroom-shaped; autozooids are more numerous and smaller at the margin.

The autozooids are apparently incapable of complete retraction. Each is covered by a thick coating of spicules. The tentacles are long (from 4–5 mm.) and laterally compressed, with the pinnules on either side of the middle line of the oral surface, thus leaving a broad free space on the aboral surface. Several of the autozooids reach a length of 18 mm., but on the margin many are only 3·5–4 mm. in length.

Between the autozooids the surface of the capitulum is closely covered by small raised points which mark the positions of the siphonozooids.

In the autozooids the spicules are (1) long rods and spindles with a few very small spines, (2) small club-like spicules with a number of prominent spines at one end and the other end tapering to a point, and (3) small double clubs. The following measurements were taken of length and breadth in millimetres :—

1. 0·25 × 0·02 ; 0·3 × 0·02 ; 0·5 × 0·02.
2. 0·12 × 0·02 ; 0·12 × 0·04 ; 0·14 × 0·04.
3. 0·06 × 0·05 ; 0·07 × 0·04.

The heads of the second group of spicules lie close together and the projecting spines interlock, thus forming a matted felt-work.

Locality : Station 232 ; 7° 17′ 30″ N., 76° 54′ 30″ E. ; 430 fathoms.

The second, and at first very puzzling specimen, is a beautiful colony of a reddish colour, surrounding a siliceous rod, which is from 2–3 mm. in breadth and 300 mm. in length. The average breadth of the colony, polyps included, is 22 mm., of the cœnenchyma, apart from the polyps, 8–12 mm. At the basal end of the colony the rod is left bare for 50 mm. There can be little doubt that this rod, which serves as an extrinsic support to the colony, is an immense sponge spicule, such as has been described in *Monorhaphis.*

The colony consists of about 75 large slightly-retractile autozooids, 10–12 mm. in height, about 4 mm. in diameter, and with tentacles 3–5 mm. in length. These autozooids arise from the finely granular cœnenchyma, apparently without regular arrangement, at intervals of 2–6 mm., and on all sides. The spaces between them are covered with very numerous siphonozooids.

Extending in a spiral up the siliceous rod there is a broad (2·5–4 mm.) band of thin cœnenchyma, which bears neither autozooids nor siphonozooids. At six places the opposite margins of thick polyp-bearing cœnenchyma arch over the thin cœnenchyma and come into contact; at two other places the thick cœnenchyma completely surrounds the rod. It may be suggested that the broad bare band corresponds to the stalk.

Along each margin of the thick cœnenchyma as it abuts on the bare tract, there is a row of minute autozooids (about 1 mm. in diameter) with brownish tentacles, and between these and the typical autozooids there are many intermediate sizes. The minute marginal autozooids are separated by intervals, varying from 1–2·5 mm.

The surface of the cœnenchyma appears to be covered by minute spiculated warts, which mark the apertures of the siphonozooids. Between the cœnenchyma and the siliceous rod there is a thin film of faintly yellowish débris.

The autozooids have a much wrinkled surface, due to contraction. The tentacles are simply infolded, and bear 12–15 plump conical pinnules. The tentacles have large cavities flattened in the radial plane, their aboral surface (in the preserved state, at least) is a narrow ridge bearing longitudinal rows of spicules. In one case a tentacle was seen to be completely inturned into the stomodæum. The wall of the stomodæum is substantial, and transversely annulated, recalling a similar appearance in *Sarcophytum agaricoides.* It is continued inwards below the surface of the cœnenchyma, almost as far as the siliceous rod.

A slice parallel to the surface of the cœnenchyma exposes the crowded cavities of the siphonozooids, usually about 0·75 – 1 mm. in diameter, and are at once seen to be crowded with ova (0·3 mm. in diameter) and blastula *embryos* (0·5 mm. in diameter). In some cases gastrulation had occurred.

A section in the plane of the longitudinal axis of the minute marginal autozooids at one of those regions where the opposite margins of thick cœnenchyma overlap the bare band, shows elongated cavities leading from the polyp openings into the cœnenchyma proper. It may also be seen that minute canals cross the band of thin cœnenchyma from side to side.

The spicules agree closely with those of the other specimen. The following measurements were taken of length and breadth in millimetres:

(a) Cœnenchyma: (1) Long spindles, with few blunt spines, 0·5 × 0·04; 0·4 × 0·03; 0·3 × 0·03.
(2) Shorter warty spindles, 0·25 × 0·03; 0·2 × 0·02.
(3) Rods with warty ends and warts in two whorls, 0·14 × 0·06; 0·12 × 0·06.
Some of these approach double clubs, while others are almost stellate.
(4) A few crosses also occur, 0·2 × 0·2; 0·2 × 0·1.

(b) Autozooids: (1) Long spindles, as in cœnenchyma, but with fewer spines, 0·55 × 0·03; 0·45 × 0·03.
(2) Clubs with bare shaft and warty end, 0·15 in length, shaft 0·02 broad, end 0·03 broad; 0·1 × 0·015 × 0·02.
(3) Short warty rods, as in cœnenchyma, 0·1 × 0·08; 0·1 × 0·06.
(4) A few crosses, 0·2 × 0·1.

Locality: Station 254; 11° 16′ 30″ N., 92° 58′ E.; 669 fathoms.

Sarcophytum agaricoides, n. sp.[1] Plate I. fig. 3.

The general appearance of the colony may be described as mushroom-shaped. A distinct cylindrical stalk expands into a large hemispherical lobe or pileus. The colour is a uniform purplish-red except on the retractile portions of the autozooids, which are yellowish-white. The pileus is approximately circular in outline, markedly convex on its upper surface and slightly concave on its lower. The total height of the colony is 5·5 cms. The pileus has a maximum breadth of 3·2 cms. and a length of 1·2 cm.

The lower surface of the pileus is covered with an epidermis, continuous with that of the stalk, and totally devoid of polyps. The upper surface bears numerous retractile autozooids, uniformly distributed over the surface. In the inter-spaces a great number of small whitish spots mark the positions of the siphonozooids. The stalk is cylindrical, and has a maximum length of 3·3 cms.

A longitudinal section through the colony shows a series of tubular canals, running from the apertures of the autozooids, bounded by abundant connective

[1] This beautiful species was described by one of my students, Mr. James Hector, as an exercise for the B.Sc. degree.—J. A. T.

tissue supported by numerous elongated rod-like spicules. These canals gradually converge towards the centre of the pileus, finally forming a series of straight parallel canals running down the stalk.

The spicules are very uniform throughout, the main differences being in the relative size.

The normal type is rod-like, usually straight, occasionally slightly bent, tapering to a point at both ends. The surface is smooth, or covered with slight projections most numerous towards the ends. Besides these, there are much smaller, very irregular spicules. In the outer tissue of the pileus there are long spicules with a few of the irregular type; in the deeper tissue of the pileus they are similar but slightly larger. In the stalk the simple and the irregular forms are almost equally well represented. In the tentacles the spicules are smaller and very numerous. A third type is also found, intermediate between the others already referred to, namely a short cylinder, truncated at one end and presenting a knobbed appearance at the other.

The autozooids are circular in outline, and may be almost completely retracted within their respective cavities, the edges of which are distinctly tumid. The apertures are about 3·5 mm. across; the diameters of the canals decrease as they pass inwards.

At the apex there are eight tentacles of the usual type bordering the mouth aperture. The anthocodial part consists of a yellowish tube with the surface much wrinkled. Its outer wall is continuous with the epithelium of the surface of the pileus, and at the level of the pileus forms a semi-transparent flexible membrane closing the entrance.

The stomodæum is continued about 10 mm. below the surface of the pileus, and is attached to the wall by the mesenteries, which bear gastric filaments, but show no trace of gonads. The stomodæal wall is very substantial, and supported by reddish spicules; it shows very conspicuous and regular transverse corrugations—27 in number; the siphonoglyph is broad and deep; the inferior opening of the stomodæum into the cœlenteron resembles a flattened papilla.

The apertures of the siphonozooid cavities are very minute, and are seen on the surface of the pileus as a large number of white spots. Each spot is a delicate membrane roofing the cavity of the polyp, and perforated by an elongated slit, the mouth of the zooid. Tentacles are unrepresented. The mouth leads into a small chamber separated from the wall of the cavity by a narrow space crossed by mesenteries.

The siphonozooid cavities pass gradually into the autozooid cavities; a number seem to be continued into the stalk as distinct channels. Others seem to end blindly. Numerous ova occur on the walls of the siphonozooid cavities a very short distance down the tube.

Locality: Station 204; 6° 50′ 20″ N., 79° 36′ 20″ E.; 180–217 fathoms.

Family *Nephthyidæ*.

This family is represented by the following forms:

Subfamily Spongodinæ:

Spongodes uliginosa, n. sp.

„ *alcocki*, n. sp.

Lithophytum indicum, n. sp.

Subfamily Siphonogorginæ:

Chironephthya variabilis, Hickson.

„ *macrospiculata*, n. sp.

Stereacanthia indica, n.g. et sp.

Agaricoides alcocki, Simpson, n. g. et sp.

Subfamily Spongodinæ.

Spongodes uliginosa, n. sp.

This species belongs to the section Divaricatæ and to the *Suensoni* group as defined by Kükenthal.

The colony is small, 21 mm. in maximum height and 28 mm. in width. It resembles a thorny bush, and is greyish-white in colour. The trunk is short, and the first branches arise at a height of 10 mm. from the base. The lowest branches are flat and reflexed, forming a complete collar round the upper part of the trunk.

The polyps arise singly on short twigs. The twig is composed of large spicules which are placed longitudinally and project beyond the polyp base, thus giving rise to the thorny appearance. The polyp heads are borne on short peduncles, and the two together reach a height of 3 mm. On the anthocodia the spicules are arranged in a crown and points; in the crown there are two to three transverse rows, and in each point there are two prominent converging spicules. These separate a little at their free ends and form a circle of projecting spines round the retracted tentacles. On the aboral surface of the tentacles there are bands of spicules arranged longitudinally. The spicules of the general cœnenchyma are spindles either straight or twisted, and covered with numerous warty projections. Their measurements, length by breadth in millimetres, are as follows:

$8 \times 0{\cdot}5$; $7 \times 0{\cdot}5$; $5{\cdot}5 \times 0{\cdot}35$; $2 \times 0{\cdot}3$; $0{\cdot}6 \times 0{\cdot}05$.

Some of the spindles are bifid at one end, or give off a short projection from one of the lateral surfaces.

The polyp spicules are warty spindles and irregular rods. They may be straight, curved, or sharply bent at one end. The following measurements were taken of length and breadth in millimetres:

$1{\cdot}6 \times 0{\cdot}1$; $1{\cdot}5 \times 0{\cdot}1$; $0{\cdot}3 \times 0{\cdot}045$; $0{\cdot}3 \times 0{\cdot}02$; $0{\cdot}16 \times 0{\cdot}025$; $0{\cdot}1 \times 0{\cdot}03$; $0{\cdot}07 \times 0{\cdot}02$.

This species comes very near *Spongodes suensoni*, Holm, but differs from it in the arrangement of the polyps and in the large dimensions of the spicules. The colour of the colony is also quite different, being in this case greyish-white, while *Spongodes suensoni* is greyish-yellow with yellow or red polyps and red "Stützbündel" and polyp spicules.

Locality: Station 237; 13° 17′ N., 93° 07′ E.; 90 fathoms.

Spongodes alcocki, n. sp. Plate I. fig. 4; Plate VIII. fig. 6.

This is a divaricate form, and belongs to the *Cervicornis* group of Kükenthal.

The base of the trunk forms a flattened disc of attachment. About 12 mm. from the base the trunk is surrounded by an irregular, flattened branch, which is interrupted altogether at one place and is twice perforated. At the perforation next the interruption there is an almost cylindrical portion which may represent a separate secondary branch. Above the collar some small branches are also slightly flattened.

The branching of the colony is on the whole in one plane, with a median and two lateral main branches, each breaking up into numerous twigs. The height is 60 mm. from above the collar, and the maximum breadth 90 mm. The colony is very delicate and readily torn; the colour is translucent white, except at the tips of the twigs and in the polyps, where the spicules are deep yellow or orange red.

The polyps occur in bundles of 5–10, but the larger numbers are the more frequent. Each polyp stands on a separate stalk, about 2·5 mm. in length. The polyps and their tentacles are white, disguised at first sight by the orange red spicules.

The spicules of the anthocodia are arranged in 8 triangles, each consisting of a pair of converging spicules and a horizontally-placed curved spicule which forms the base. In addition to these there are some irregularly horizontal spicules. The aboral surface of the tentacle has a band of minute red spicules arranged in two rows.

The spicules of the polyp stalk consist of closely contiguous obliquely transverse rows; on the surface of the stalk farthest from the polyp several stronger spicules are arranged longitudinally to form a Stützbündel, one of which projects for a short distance beyond the polyp.

The spicules of the stem form a loose network, and are spiny spindles straight or curved. The following measurements were taken of length and breadth in millimetres:

1·7×0·1; 1·4×0·1; 1·3×0·06; 0·8×0·045; 0·75×0·05; 0·3×0·02; 0·2×0·03.

The polyp spicules are straight or curved spindles. The following measurements were taken of length and breadth in millimetres:

1·5×0·09; 0·9×0·06; 0·7×0·04; 0·16×0·03; 0·12×0·03; 0·06×0·03.

The colour of the stem and branches is white, the twigs orange red, and the polyp heads appear brick red.

Locality: Bay of Bengal; 88 fathoms.

Lithophytum indicum, n. sp.

This species is represented by a small fragment, 21 mm. long and 14 mm. broad.

The basal attachment is absent, and so also is the lower part of the stem, so that the exact shape of the colony is not known. The upper part of the stem divides into three lobes or short branches of nearly equal size.

The polyps are arranged on the branches at fairly wide intervals, and in a slightly contracted condition are 2 mm. in height and 1 mm. in basal diameter. The tentacles are about 0·6 mm. in length, and have on each side about eight long slender pinnules. There are abundant ova in the lower parts of the polyps. The spicules of the polyp lie transversely at the base, but just below the tentacles they are arranged in eight longitudinal rows.

The polyp spicules are spindle-shaped, varying in length from 0·15–0·5 mm. and in breadth from 0·02–0·08 mm. They are covered by sparse fairly strong simple spines. A few are slightly thicker in the middle, and marked by an X-shaped mark near the middle point.

The tentacle spicules are shorter, mostly about 0·1 mm. in length.

The spicules of the stem are blunt spindles or rods covered with sparse simple spines, and a few irregular incipient quadriradiate forms. They vary in length from 0·2–0·9 mm. and from 0·03–0·06 mm. in breadth. The quadriradiate forms vary in length from 0·2–0·4 mm., and are all marked by an X-shaped mark at the origin of the rays; sometimes one of the cross-marks is so faint as to seem to be absent.

This species is probably to be referred to the vicinity of *L. africanum*; but it is distinguished by its large polyps, large spicules, and the presence of the quadriradiate forms.

Locality: Station 333; 6° 31′ N., 79° 38′ 45″ E.; 401 fathoms.

Subfamily Siphonogorginæ.

Chironephthya variabilis, Hickson.

This species is represented by a large number of fragments, all of which probably belonged to one large colony. They present a rather striking appearance owing to the contrast between the general white or very pale pink colour of the branches and the deep coral red colour of the spicules of the anthocodiæ.

The spicules of the anthocodiæ are arranged in two groups, the crown and point spicules, the point spicules being arranged *en chevron* and not in a *fan-shaped* manner.

The measurements of the spicules are as follows:

Point spicules, 0·35×0·025; 0·3×0·03; 0·3×0·025; 0·2×0·02.

Crown spicules, 0·6×0·035; 0·5×0·028; 0·4×0·03; 0·3×0·02.

On the aboral surface of the tentacles there are numerous small spicules arranged in Vs. Their average length is 0·08 mm., and their average breadth 0·012.

Locality: Bay of Bengal; 88 fathoms. Previously recorded from the Maldives (Hickson); Ceylon (Thomson and Henderson, 1905).

Chironephthya macrospiculata, n. sp. Plate IV. fig. 10.

This species is represented by a single colony about 5 cms. in height.

From the flat spreading base two stems arise which are fused together for a short distance. The larger gives off a lateral branch at a point nearly half-way up. The colony presents a very rough appearance owing to the huge spicules which form a loose covering over the stems and branch. The spaces left between the large spicules are filled by minute spicules which are either colourless or red.

The polyps arise all round the stems and branch, and are supported by small platform-shaped calyces formed by several spicules placed side by side. In the polyps the spicules are arranged in crown and points. The crown consists of three rows, and in the points the spicules are arranged *en chevron*. In each point there are two principal spicules slightly inclined to one another, and two smaller spicules in the angle between these. Two pairs lie in the space between two adjacent points. On the aboral surface of each tentacle there is a band of spicules arranged longitudinally or in Vs, with the apices of the Vs directed towards the distal end of the tentacle. The spicules of this band are white, except for a narrow transverse portion about half-way up in which the spicules are a bright red.

The spicules of the cœnenchyma are warty or spiny spindles, straight or slightly twisted, coloured or colourless. The following measurements were taken of length and breadth in millimetres:

1. Uncoloured, 8·3×0·9; 5·2×0·9; 3·9×0·6; 1·4×0·22; 0·8×0·12.
2. Coloured, 0·3×0·04; 0·2×0·02; 0·15×0·02.

The polyp spicules are warty spindles, either straight, curved, or sharply bent at one end. They are colourless, with the exception of those composing the red transverse band on the tentacles. The following measurements were taken of length and breadth in millimetres:

0·8×0·1; 0·8×0·08; 0·6×0·1; 0·2×0·03.

The colour of the colony is yellowish-brown.

This species is easily distinguished from previously described forms by the immense size of the spicules in the cœnenchyma.

Whitelegge described *Siphonogorgia macrospina* from Funafuti with spicules sometimes 6 mm. in length, and Miss Hiles has described an apparently similar form in which the largest spicules were 4·4 × 0·37 mm. (MS. kindly lent by Professor Hickson). Through Professor Hickson's courtesy we have been able to examine this specimen, and we find that it is very different from ours.

Locality: Station 246; 11° 14′ 30″ N., 74° 57′ 15″ E.; 68–148 fathoms.

Stereacanthia indica, n. g. et sp. Plate V. fig. 2; Plate IX. fig. 19.

This new genus is represented by two broken specimens, of which the more complete, when pieced together, is about 11 cms. in height.

The colony consists of two parts: (1) a long bare trunk, and (2) the branched polyp-bearing portion. The trunk has a thickened basal portion in which there are many projecting spicules. It is thin-walled, and consists of a number of longitudinal canals with the adjacent walls fused and containing numerous large spicules which help to give rigidity to the stalk. The polyp-bearing portion consists of two or three branches irregular in shape and of various lengths, with the polyps closely disposed. The general colour of the polyparium is a light brown, while the long trunk is white, slightly brownish at the base.

The trunk is composed of about 20 canals with thin walls fused together and filled with spicules. Owing to the contraction of the preserved specimens and the thinness of the partition walls the stalk or trunk is greatly shrivelled and somewhat broken, and thus the mode of growth is not at all clear; but it seems that the polyp cavities are continued down and come into connection with the canals in the branches, which are either new interstitial growths or prolongations of the canals of the trunk.

The trunk is divided into two irregular branches, which again divide into a number of smaller branchlets or lobes, thus giving the polyp-bearing part the appearance of a rugged irregular bush. The branches are unequal in size, and in one specimen the thicker of the two main branches divides almost at its origin into two principal parts.

The polyps occur both on the primary branches and on the secondary branchlets or lobes. On the main branches they are scarce and arise singly, scattered over the whole surface, but usually leaving a bare strip on the outer surface of the branch. On the branchlets they are numerous and crowded.

The polyps are arranged either singly or in small groups of about seven, and are borne on stalks about 2 mm. in length, the basal portions of which may be in close contact. They stand at an angle to the stalks. On the polyp stalks the spicules are arranged obliquely transverse, except that on the dorsal side several slightly stronger spicules are arranged more regularly in a longitudinal direction, but they are not sufficiently marked to warrant their being classed as Stützbündel spicules. On the anthocodial cup the spicules are arranged in eight irregular triangles rising from a ring of transverse spicules, and thus recalling the arrangement of the spicules in the anthocodiæ of *Chironephthya*. On the aboral surface of each tentacle there is a band of longitudinally placed spicules, and these bands, when the tentacles are folded in, form a low eight-rayed rudimentary operculum.

The spicules of the outer wall of the stalk are large spindles. In the lower part near the base they lie irregularly and somewhat transversely, while in the upper part they become more regular and are placed longitudinally.

The spicules of the outer wall of the stalk are straight or curved spindles, thickly covered with very rough prominent warts. The measurements of length and breadth in millimetres are:

3·4×0·35; 2·2×0·25; 1·9×0·22; 1×0·15.

The spicules of the canal walls in the lower part of the stalk are very large spindles, thickly covered with very rough warts, and either straight, curved, or twisted. The measurements of length and breadth in millimetres are:

7×0·7; 7×0·6; 6·5×0·8; 3·3×0·5; 3×0·6; 0·7×0·2; 0·4×0·1.

The spicules of the polyps and polyp stalks are warty spindles either straight or curved, or somewhat "golf-club" shaped; the warts are prominent and rough. The smaller spicules are freer from projections, but some have almost serrate edges. Their measurements are:—

1·8×0·175; 1·5×0·2; 1·2×0·15; 1×0·15; 0·4×0·03; 0·12×0·02; 0·1×0·02.

Systematic Position.—The infolding of the tentacles when at rest, the presence of the large canals in the main stem and branches, the thick external layer of the main stem, show that this form belongs to the Nephthyidæ. Adopting Kükenthal's revised classification of this family, the specimen must be placed, owing to the absence of Stützbündel spicules, in his first division. In this division he included, A, those with the "canal walls not thickly filled with spicules," and B, those with the "canal walls thickly filled with spicules." As there are abundant spicules in the canal walls of our specimen, it must be placed in the B. group and in the second section of this group, which is characterised by the polyps occurring singly or in bundles. In this section there are two genera, distinguished by the absence or presence of an irregular internal axis built up of spicules closely packed together. As the present specimen has no trace of an internal axis, it must be placed near the genus *Lemnalia*. From this it differs in several respects, *e.g.* in having no "quadriradiate double-stars" in the wall of the trunk or stalk. It seems necessary to refer it to a new genus, *Stereacanthia*, which may be thus defined:

Colony upright, consisting of two parts: (1) a bare densely spiculose trunk composed of large thin-walled longitudinal canals, with fused walls but with no trace of an irregular central axis; and (2) a branched polyp-bearing portion or polyparium, on which the polyps are arranged singly, or in small bundles crowded together on small branchlets. The tentacles during rest are not retracted, but simply infolded, the aboral bands of tentacular spicules thus forming a pseudo-operculum. The spicules are covered by rough prominent warts, and are mostly straight or curved spindles. A number of polyp spicules are "golf-club" shaped, *i.e.* are bent at a sharp angle near one end.

Locality: Eight miles west of Interview Islands, Andamans; 270–45 fathoms.

In another haul from near the Andamans a young colony of this species is included. It rises from a flat base and consists of two stems—3·2 cms. and 3 cms. respectively in height. In this specimen the main stems do not break up into branches, but give off small twigs. They are slightly flattened, and polyps occur singly over their whole surface.

Locality: Andamans; 270–45 fathoms.

Agaricoides alcocki,[1] Simpson, n. g. et sp. Plate VIII. fig. 4; Plate X. figs. 1–19.

This new genus is represented by numerous specimens varying greatly in size, which illustrate different stages of growth.

The following are the measurements of some of the more perfect specimens:

Total height.	Maximum breadth.	Length of Trunk.	Thickness of Trunk.
3 cms.	3 cms.	1·75 cms.	1·5 cms.
2·5 ,,	3 ,,	1·5 ,,	1 ,,
3·2 ,,	1·5 ,,	2·2	0·8 ,,
0·9 ,,	1·2 ,,	0·8 ,,	0·6 ,,

All were attached to pieces of Madrepore coral, probably *Lophohelia* sp.

General Description.

The colony consists of two very distinct parts: (1) a bare trunk, and (2) a polyp-bearing "pileus." The trunk is composed of a large number of longitudinal canals with the adjacent walls fused and densely packed with spicules, which give consistency and rigidity to the colony. The upper umbrella-shaped portion or "pileus" so closely resembles a mushroom that the term "agariciform" might well be applied. The general colour of the pileus portion of the colony is a pale orange yellow, while the trunk is whitish. The zooids, whose tentacles are not retractile, are introversible within cylindrical stalks which are expanded terminally into characteristic octagonal disc-like expansions. The close-set octagons covering the

[1] For a description of this remarkable new type we are indebted to Mr. James J. Simpson, M.A., B.Sc. See *Zool. Anzeig.* xxix. (1905) pp. 263–271, 19 figs.

pileus, with a zooid appearing in the centre of each, present a very remarkable and unique appearance.

The Canal System.

Owing to the contraction of the spirit-specimens, the mode of growth is by no means obvious, but an examination of various stages, and of serial sections through these, makes the mode of increase in the number of the canals fairly clear. The centre of the stem is occupied by a number of large canals whose cavities are about 3 mm. in diameter. These do not communicate with one another, but at several points solenia can be seen connecting them with the canals in the cortical region whose cavities vary from 2 mm.–1 mm. in diameter. The cortical canals terminate basally in a cul-de-sac, while upwards they increase in diameter so as to give rise to zooids on the pileus portion. The younger zooids are peripheral, and the whole colony may thus be compared to a bundle of compound racemes, the branches of which are hollow, and where the secondaries and tertiaries fuse to the primaries and grow to an equal length with them, so as to result in a corymb-like expansion. The eight mesenteries of the zooids are continued downwards almost to the very base of the canals, and at the same time the asulcar pair can be clearly distinguished by the characteristic ciliated groove. This is also the case in *Siphonogorgia* and *Lemnalia*, while in some of the Nephthyidæ, *e.g. Spongodes*, only the asulcar mesenteries are continued into the canals of the stem.

Origin of the Zooids.

Both the central or primary canals and the cortical or secondary canals give rise to zooids in a remarkable and interesting manner. After attaining a certain height, which is practically uniform for the colony, the walls turn inwards, so that the cavity is thus reduced in diameter; and when this is approximately one-half of the original measurement a vertical upgrowth again commences, thus forming a cylindrical cup-shaped projection, homologous to the verruca in the Axifera. When the height of this part is about 4 mm. the circumference grows out into eight digitiform structures, while the wall again growing inwards fuses with the lower part at the eight indentations, forming a similar number of short blind tubes. This constitutes what might be termed the verruca proper. Growth still takes place, and a zooid is the result, consisting of a comparatively long stalk bearing the anthocodia. That this is the mode of growth is clearly demonstrable in the younger colonies, and also in the less advanced polyps round the periphery of the older colonies. As the colonies grow in size, the verrucæ also become more complicated, the terminal stellate part expanding horizontally to form an octagonal disc, with the indentation less pronounced, containing eight cavities which correspond to those formed by the retractor muscles. To complicate matters still further, towards the centre of the older colonies, verrucæ which correspond to the primary canals fuse with the adjoining verrucæ so that the canals are now continuous.

The partitions of the canals are densely spiculose, in addition to the outer felt-work, and a cross section shows that the spicules are arranged for the most part vertically, the cavities conforming to the tubercles on the spicules.

Structure of the Zooids.

The zooids are about 5 mm. in height, and consist of a somewhat slender stalk bearing a cup-like anthocodia, the whole being densely covered with a felt-work of minute warty spindles. The tentacles are short and broad, with a single row of pinnules on each side; their bases are confluent, so as to enclose a capacious hollow —the oral disc, over which they can be infolded. They are not retractile, but when at rest, being infolded, the biserial arrangement of the spicules forms a very primitive operculum.

The wall of the anthocodia is prolonged into eight triangular lobes, on which the spicules are also arranged biserially, so that each pair forms two sides of a triangle, the enclosed angle becoming more and more obtuse towards the base. This arrangement is continued down to the origin of the stalk, forming a series of ridges on the anthocodia. The triangular projections thus function as a protection to the infolded tentacles. The stalk, though narrow, is very elastic, because the zooid when at rest forms an introvert within it, which in turn sinks within the verruca. The zooid is withdrawn by eight strong bands of retractor muscles, which thus form eight cavities running upwards, and corresponding to the canals in the octagonal disc. These retractor muscles pass downwards and are continuous with the eight mesenteries of the zooid.

The oral disc is spacious and circular, containing a rather large elliptical mouth-opening, which leads into a keyhole-shaped, richly-ciliated stomodæum, in which a very distinct sulcus can be distinguished. The ectodermic cells in this region are more numerous and the cilia are longer.

The mesenteries are all complete, and the muscle banners on the sulcar aspect are easily discernible. The filaments are continued down the stem canals almost to the very base of the colony, while the asulcar filaments show very markedly the ciliated groove so characteristic of the group.

Ova of enormous size are present in great abundance attached to the mesenterial filaments. They vary from 0·1 mm.–0·6 mm. in diameter, but although a considerable number of the larger and more mature ova were stained with borax-carmine, no sign of segmentation could be found, so that the question of viviparity for this genus must remain undecided.

A fact which may prove to be of great interest is that in every zooid examined there was a large number of Foraminifera of various kinds, and in the decalcified sections examined the protoplasmic contents could be seen surrounded by the ectodermic cells of the stomodæum, while many were also enclosed within the pinnules of the tentacles. When the zooids are at rest, the tentacles are infolded; and as many

Foraminifera are enclosed by the pinnules, they must have entered while the polyp was expanded. The absence of food in the cœlentera of most Alcyonarians and the frequent presence of zoochlorellæ point to the fact that many Alcyonaria are symbiotic organisms, but the fact we have noticed suggests that some have the power of assimilating food from other sources.

Spicules.

The great majority of the spicules are arranged irregularly so as to form a dense felt-work on the surface of the canals, giving consistency and rigidity to the whole colony; but many are also embedded in the mesoglœa. On several parts of the colony, however, the arrangement is particularly regular, *e.g.* on the expanded disc of the verruca, on the protective prolongations of the wall of the anthocodia, and also on the tentacles, where in all cases the arrangement is biserial.

All are irregularly echinate, while many may be called warty. The spines vary greatly in form and size, some sharp and triangular standing at right angles to the spicules, others hooked and thorn-like, while others are truncated. They are, for the most part, simple, but compound forms are not infrequent. No type, however, can be said to be characteristic of any particular part of the colony.

The spicules consist chiefly of straight and curved spiny spindles, some approaching the "scaphoid" form so characteristic of the genus *Gorgonia*, thus showing convergence in another direction. Other forms are single clubs and clubs with a curved termination, resembling "hockey-clubs."

In the decalcified sections an organic residue was to be seen in the spicule cavities in the mesoglœa, and an examination of the smaller and more transparent spicules showed that there was an organic axis with branches which passed out into the spiny projections.

It is interesting to note that the same has been recorded by Bourne for *Lemnalia*, the genus most closely related to our form.

The following measurements of spicules in millimetres were taken:

a. Transparent spicules of the outer wall of the trunk.

Straight spindles, $0{\cdot}85 \times 0{\cdot}03$; $0{\cdot}75 \times 0{\cdot}04$.

Curved spindles, $0{\cdot}7 \times 0{\cdot}025$; $0{\cdot}75 \times 0{\cdot}03$.

b. Transparent spicules of the partition walls.

These show a greater preponderance of straight forms and bear more compound spines.

Spindles, $0{\cdot}85 \times 0{\cdot}05$.

Single clubs, $0{\cdot}7 \times 0{\cdot}03$.

c. Transparent and pale yellow spicules of the disc-like expanded portion of the verruca.

On the whole these are smaller than the two preceding groups and hardly so spinose.

Spindles, $0{\cdot}65 \times 0{\cdot}02$; $0{\cdot}6 \times 0{\cdot}01$.

d. Spicules of the anthocodia mostly pale yellow with few and small spines.

Straight spindles, 0·6 × 0·015 ; 0·4 × 0·02.

Curved spindles, 0·55 × 0·015 ; 0·55 × 0·01.

Hockey-clubs, 0·6 × 0·2 ; 0·4 × 0·015.

e. Pale yellow spicules of the tentacles mostly curved spindles and hockey-clubs.

Curved spindles, 0·45 × 0·02 ; 0·35 × 0·03.

Hockey-clubs, 0·35 × 0·04.

Systematic Position.

The presence of the close felt-work of spicules already referred to places this form in the subfamily Siphonogorginæ as defined by Wright and Studer. The genera *Siphonogorgia* (Kölliker), *Chironephthya* (Wright and Studer), *Paranephthya* (Wright and Studer), and *Scleronephthya* (Wright and Studer), need not be considered ; but there is undoubtedly relationship with *Lemnalia* (Gray emend. Bourne). From this genus, however, our specimen differs essentially, in that it is not branched, and in having the anthocodiæ pedicelled. Other features, such as the form of the verrucæ, the nature of the anthocodiæ, the introversion of the zooids, and the general details of the colony, mark it off as a new and very distinct genus.

Diagnosis.

Colony upright, attached, mushroom-shaped (agariciform), consisting of (1) a stout, densely spiculose trunk composed of a longitudinally arranged system of cœnenchymal canals with fused walls, the more superficial of which appear as ribs on the surface ; and (2) a "pileus" portion bearing the zooids which are introversible within projecting verrucæ—cylindrical extensions of the trunk canals, the upper portions of which are expanded peripherally into octagonal discs containing eight canals, corresponding to the eight compartments formed by the retractor muscles. The anthocodiæ are borne on somewhat slender stalks, the elastic walls of which are continuations of the upper walls of the discs. The tentacles are not retractile, but are simply folded over the wide oral disc, the biserial arrangement of the spicules forming a simple pseudo-operculum. The oral disc is spacious, protected by eight triangular projections of spicules. The mouth, considerably elongated, leads into a richly ciliated stomodæum in which a distinct sulcus can be distinguished. The mesenteries are complete, and are continued down to the very base of the stem canals. The spicules are irregularly echinate, and consist chiefly of straight and curved spindles ; while some approach the "scaphoid" type, others are single "hockey-clubs," *i.e.* club-shaped with a curved termination.

Locality : Station 333 ; 6° 31′ N., 79° 33′ 45″ E. ; 401 fathoms.

ORDER III. PSEUDAXONIA, G. von Koch.

Family *BRIAREIDÆ*.

Subfamily Briareinæ.

Paragorgia splendens, n. sp.

Family *SCLEROGORGIDÆ*:

Suberogorgia köllikeri, Wright and Studer, var. *ceylonensis*, Thomson.

Keroëides koreni, Wright and Studer.

,, *gracilis*, Whitelegge.

Family *MELITODIDÆ*:

Parisis indica, n. sp.

Family *CORALLIDÆ*:

Pleurocorallium variabile, n. sp.

Family *BRIAREIDÆ*.

Subfamily Briareinæ.

Paragorgia splendens, n. sp. Plate I. fig. 5; Plate V. figs. 9 and 14.

A beautiful coral-red colony 12·5 cms. in height and 6·5 cms. in maximum breadth represents this species. The branching is confined to one plane, and the polyps are directed mainly towards one aspect. The diameter of the main stem near the base is 8 mm. The pseudaxial portion consists of long warty spindles tightly bound together, and is penetrated by numerous solenia. In the centre of the mass there are five main canals, four larger surrounding a smaller. The cœnenchyma is moderately thick, and has a glistening arenaceous appearance. In the older part of the stem it is smooth, but on the younger portions and on the branches it is marked by irregular ridges and furrows which extend into the verrucæ. The polyps may occur singly or in groups, the tips of the branches being swollen into large club-shaped heads containing sometimes nine or ten polyps. They show the characteristic furrowing of the cœnenchyma, and when inturned present an eight-rayed structure. The tentacles are of a deeper red colour than the general cœnenchyma. The colony is dimorphic, numerous siphonozooids occurring over the whole cortex. This phenomenon is apparently absent in some of the other fragments. No trace of tentacles could be found in the siphonozooids, but the characteristic Alcyonarian internal structure is plainly visible, the muscle-banners being most developed on the directive mesenteries.

The spicules of the axis (a) are typically spindle-shaped, irregularly covered with markedly projecting warts or spines. Many have broad bifurcated ends. All are pink or colourless.

Those of the cortex (*b*) are of a salmon-pink colour. They are shorter and more closely covered with large warts, and some resemble a dumb-bell with a very short handle. The following measurements of length by breadth in millimetres were taken :

(*a*) 0·22×0·045 ; 0·2×0·04 ; 0·18×0·04 ; 0·15×0·03.
(*b*) Spiny, 0·14×0·02 ; 0·12×0·03 ; 0·1×0·025.
Warty, 0·12×0·03 ; 0·1×0·03 ; 0·06×0·025.

We cannot refer this form either to the precisely described and figured *P. nodosa* of Koren and Danielssen, or to the vaguely described and doubtless different *P. nodosa* of Gray.

Locality : Station 284 ; 7° 55′ N., 81° 47′ E. ; 506 fathoms.

Family *Sclerogorgidæ.*

Suberogorgia köllikeri, Wright and Studer, var. **ceylonensis,** Thomson.

Plate IV. figs. 11 and 12.

This species is represented by two small yellowish-brown fragments, 5·5 cms. and 5·7 cms. respectively in height, without basal attachment.

The fragments are branched in one plane, the branches arising almost at right angles to the stem and then turning upwards and running roughly parallel with it.

The axis is almost cylindrical, with a diameter of about 3 mm. in its thickest part, and 0·9 mm. at the tips of the branches. In texture it is sclerogorgic, and is marked on opposite sides by two shallow winding grooves for the nutrient canals.

On the stem the polyps arise without any regular arrangement, anywhere except on the grooves ; on the branches they are arranged in two winding rows, one on each lateral face. They are capable of complete retraction within the verrucæ, which are low truncated cones varying from 0·9–1·5 mm. in diameter. The aboral surface of the tentacles bears spicules arranged *en chevron* in the lower part, but parallel to the length of the tentacle near the tip. These form an eight-rayed pseudo-operculum when the tentacles are withdrawn.

The general cœnenchyma is thin and densely packed with spicules.

The spicules are spindles, which are sometimes almost oval in shape, and some small quadriradiate forms. All are covered by warty protuberances. The following measurements were taken of length and breadth in millimetres :

The large warty spindles, 0·22×0·06 ; 0·2×0·06 ; 0·2×0·04 ; 0·14×0·08 ; 0·11×0·1.

The small warty spindles, 0·14×0·03 ; 0·13×0·04.

The spicules of the tentacles, 0·15×0·035 ; 0·13×0·03.

The quadriradiate forms, 0·08×0·06 ; 0·075×0·05.

The quadriradiate forms are all marked by an X-shaped mark in the centre.

This specimen differs from the typical *Suberogorgia köllikeri* in having its branching almost completely in one plane, in the arrangement of the polyps on the smaller branches, and in the smaller size of the spicules.

Locality: Andamans; 270–45 fathoms. Previously recorded from Ceylon and Zanzibar.

Keroëides koreni, Wright and Studer. Plate I. figs. 6 and 7.

A single specimen was obtained from the neighbourhood of the Andamans. It measures 3·3 cms. in height, and has a maximum breadth of 2·9 cms. The branches are few in number, and the white polyp-heads present a beautiful contrast with the bright vermilion red of the branches.

Locality: Andamans; 270–45 fathoms.

Previously recorded from: "Challenger" Station 232, Hyalonema-ground, off Japan; depth, 345 fathoms. Also from Funafuti (Hiles); 40–90 fathoms.

Keroëides gracilis, Whitelegge. Plate IV. figs. 1, 2, and 3.

To this species we refer a number of fragments from the vicinity of the Andamans.

As far as can be made out, the colony is branched in one plane, and has a flat spreading basal attachment. The branches are given off for the most part alternately, but this is not strictly adhered to.

The axis is sclerogorgic, rigid, and dense, almost cylindrical in shape, without any distinct traces of grooving. It is brownish-yellow in the older parts, becoming paler in the younger.

The polyps arise irregularly on the main stem and the larger branches, tending, however, to become arranged alternately on the smaller branches. They are capable of complete retraction within the verrucæ, and the tentacular spicules form a low eight-rayed operculum. On the aboral surface of the tentacles the spicules are arranged with their long axes parallel to the length of the tentacle. The distance between adjacent verrucæ may be 3·5 mm., or they may be touching one another.

The verrucæ are cylindrical, composed of warty spicules arranged longitudinally, and vary in height from 1–2 mm., with a basal diameter of about 1 mm.

The general cœnenchyma is thin, and consists of one layer of warty spicules, which are arranged very irregularly, with small, brightly-coloured spicules filling the spaces between. In the older parts of the stem and branches they are small flat irregular discs arranged transversely, with here and there, especially near the origin of a branch, a number of long flat spindles arranged longitudinally. In the younger parts of the branches this arrangement is replaced by one in which the long spindles predominate. These are longitudinally arranged, and very few discs are to be seen.

The following measurements were taken of length and breadth in millimetres: 2·2×0·35; 1·4×0·35; 0·6×0·2; 0·5×0·25; 0·4×0·22; 0·4×0·35; 0·35×0·1.

The tentacle spicules are 0·11–0·15 in length by 0·02–0·03 in breadth.

This form comes very near *Keroëides gracilis*, but the following differences may be noted:

Keroëides gracilis, Whitelegge.	*Keroëides gracilis* (Indian Ocean form).
Branches alternate.	Branches not strictly alternate.
Polyps alternate.	Polyps irregular on main stem and larger branches, tending to become alternate on smaller branches.
Spicules are spindles and irregular forms; placed transversely on and in the neighbourhood of the verrucæ.	Spicules on the older parts of stem and branches very irregularly arranged; in the younger parts long spindles regularly arranged predominate.
Spicules reaching 2 mm.	Spicules 2 mm. and over.
Spicules bright brick red.	Spicules colourless or faint pink.
Colour of colony coral red.	Colour of colony pale pink, and in this respect agreeing with the specimen described by Miss Hiles.

Locality: Andamans; 270–45 fathoms.

Previously recorded from Funafuti and British New Guinea.

Family *Melitodidæ*.

Parisis indica, n. sp. Plate IV. figs. 4, 5, 8, and 9.

This species is represented by several fragments, the longest of which is 45 mm. in length.

The colony is white in colour, and much branched in one plane. The branches rise from both sides of the axis, but on the main branches the secondary branches often rise from one side only.

The axis is composed of alternate horny and calcareous joints, the latter being from 4–6 times as long as the former, and giving off the branches. Externally it is deeply grooved, and numerous solenia surround it, one in each groove. The calcareous internodes are composed of spicules closely cemented together, and in the horny nodes there are numerous small spicules. The grooves are seen on the nodes as well as on the internodes.

The polyps are arranged on each lateral surface in a sinuous row, which appears like two rows with the polyps of the one alternating with the polyps of the other. They also occur in fours, either in a spiral or in a whorl. The verrucæ are truncated cones, about 1 mm. in height, and standing about 2 mm. apart.

The surface of the colony is entirely hidden by an incrustation of sponge and *Palythoa*, which obscures the form of the calyces.

When freed from the incrustations the surface of the cœnenchyma presents a fine tessellated appearance, which is also seen on the polyp calyces. It appears thin, but this may be the effect of the incrusting sponge.

The spicules are (*a*) spindles covered with numerous rough prominent warts; (*b*) a few more oval or globular in shape; and (*c*) a few quadriradiate forms with X-shaped marking. The following measurements were taken of length and breadth in millimetres:

(*a*) Spindles, 0·2×0·075; 0·16×0·08; 0·14×0·08.

(*b*) Globular forms, 0·3×0·2; 0·2×0·1.

(*c*) Quadriradiate forms, 0·10 from tip to tip one way × 0·10 from tip to tip the other way; 0·12 from tip to tip one way × 0·08 from tip to tip the other way.

The species is near to *Parisis fruticosa*, Verrill, but is separated from it by (1) the smaller size of the calyces and of the spicules; and (2) by the pavement-like appearance of the cœnenchyma and calyces.

Locality: Andamans; 8 miles west of Interview Island; 270–45 fathoms.

Family CORALLIDÆ.

Pleurocorallium variabile, n. sp. Plate I. fig. 9; Plate V. fig. 6; Plate IX. fig. 13.

To this species we refer a number of broken fragments, the largest of which is 13·5 cms. in height and 6·2 cms. in width.

The colony is very profusely branched in one plane. The branches are tortuous, and show little, if any, sign of lateral compression. They arise from the antero-lateral surfaces of the stem, and diminish gradually in thickness towards their tips. From the sides of the large branches and of the stem, numerous short branchlets arise.

The axis is hard, not easily indented with a knife, solid, almost cylindrical in section in some parts and slightly oval in others. It is white in colour, and its surface is marked by very fine striations, often very faint.

The cœnenchyma is thin, creamy white in colour, and full of closely packed small spicules which look like glistening sand grains.

The polyps occur irregularly on the anterior surface of the stem and branches. The tentacles are about 0·5 mm. in length, yellowish in colour, and closely covered by small spicules.

The salmon-pink verrucæ present a fine contrast to the creamy white of the general cœnenchyma and the yellow of the tentacles. They are prominent, almost cylindrical in shape, marked by eight longitudinal ridges, and reaching a height of 2·7 mm. When the tentacles are retracted, the apices of the verrucæ present the appearance of eight-rayed stars.

The spicules of the cortex are of two kinds: (1) Octoradiate spicules, with a short shaft and terminal tubercles. Several tubercles project at each end of the spicule at right angles to the shaft. The large tubercles may themselves be covered with smaller tubercles. These spicules vary in length from 0·06–0·08 mm., and in width from 0·04–0·05 mm.

(2) Spicules that resemble opera-glasses in shape. They consist of two globose masses somewhat flattened at one end, and bearing at the other end short processes, variable in shape, with several tubercles. They are on an average 0·06 mm. in length, and 0·04 mm. in width.

To these may be added a very few rough spiny spindles from 0·06-0·09 mm. in length, and 0·02 mm. in diameter.

To this species we also refer a few fragments from the same locality which differ in colour, being rosy red with a yellowish-brown cœnenchyma. The arrangement of the polyps and the details of spicules and axis are, however, much the same as in the specimen described.

This species differs from *Pleurocorallium johnsoni*, Gray, (1) in being more profusely branched, and (2) in having more prominent and differently arranged verrucæ.

The spicules in this species agree in type with the descriptions given of *P. johnsoni* by Gray and Ridley; but it seems difficult to fit the specimen described by Moroff (1902) as *Pleurocorallium confusum*, n. sp., into the genus, for he says: "Coenenchym schwach mit vielen bis 0·25 mm. grossen Spicula besetzt, die eine gerade oder gekrümmte spindelförmige Gestalt aufweisen. Nicht selten sind auch plattenförmige Spicula, sowie Vierlinge zu sehen."

Locality: Station 284; 7° 55′ 00″ N., 81° 47′ 00″ E.; 506 fathoms.

ORDER IV. AXIFERA, G. von Koch.

Family *DASYGORGIIDÆ.*

Lepidogorgia verrilli, Wright and Studer.
Chrysogorgia orientalis, Versluys.
" *flexilis*, Wright and Studer.
" *dichotoma*, n. sp.
" *irregularis*, n. sp.
" *indica*, n. sp.

Lepidogorgia verrilli, Wright and Studer. Plate III. figs. 5*a* and 5*b*.

This species is represented by three fragments which apparently compose one specimen and a portion of another.

In one the basal attachment is present, and consists of a number of root-like processes very calcareous and translucent.

The axis is thin and hair-like, very calcareous and brittle, but slightly flexible near the tip.

The polyps are 3·3 mm. in height, arranged uniserially at intervals of nearly 6 mm.

The cœnenchyma is moderately thick, slightly more so on the side bearing the polyps.

On the stem the spicules are arranged side by side; on the polyp an eight-rayed pseudo-operculum is formed by spicules on the bases of the tentacles, and for a short distance below the origin of the tentacles the spicules are arranged in eight rows. The spicules are irregular in shape, curved or straight, and sometimes reaching a length of 2 mm.

Locality: Andaman Sea; 375–490 fathoms.

Previously recorded from off Japan and from Macassar Straits.

Two other specimens (*B* and *C*) agree on the whole with *A*, but differ in minor details, as shown in the following table:

Axis.	Cœnenchyma.	Polyps.
A. Square at base, cylindrical a little farther up; slender, 0·7 mm. in diameter, becoming hair-like; brittle, very calcareous, slightly flexible near the tip; iridescent; surface ridged.	Very thin, axis shining through.	At distances of 5·5 mm. Height 2–2·8 mm.

Axis.	Cœnenchyma.	Polyps.
B. Axis cylindrical throughout its whole length; slender, 0·5 mm. in diameter at lower end, then thread-like; brittle, very calcareous, slightly flexible near the tip; iridescent; surface ridged.	Slightly thicker than in *A.* Axis only faintly visible.	At distances of 6 mm. Height up to 4 mm.
C. Axis square, with the angles rounded off farther up; fairly strong, 1·15 mm. in diameter at base, becoming hair-like at tip; brittle, very calcareous; slightly flexible near the tip, golden in colour; surface ridged.	Thin, axis distinctly visible.	At distances of 5–9 mm., but in the longer intervals there is usually a young polyp between. Height up to 4 mm.

COMPARATIVE TABLE OF SPECIES OF LEPIDOGORGIA.

Species.	Shape of Axis.	Colour.	Branching.	Surface.	Texture.	Cœnenchyma.	Polyps.
L. petersi, Wright and Studer.	Irregularly four-sided, with angles rounded below.	Yellow, with marked golden lustre.	None.	Marked by numerous short grooves.	Brittle, very calcareous.	Thin, membranous, with few spicules.	At intervals of 3–3·5 mm.; may be 7 mm. in height, but usually about 4 mm., with the axial side concave; the three tentacles on the axial side smaller than the others, and the median axial tentacle rudimentary.
L. verrilli, Wright and Studer.	Cylindrical.	Yellowish-white, with slight golden lustre.	None.	Smooth.	...	Thin, with few spicules.	At intervals which vary from 4–4·7 mm., or from 2–6 mm.; usually 3–3·5 mm. in height; axial side concave with very small tentacles.
L. challengeri, Wright and Studer.	...	...	None.	Lower part as in *L. petersi*, upper portion nearly smooth.	...	Smooth, with very few spicules.	At intervals of from 4–5 mm., but may be 7 mm.; 2·5–3 mm. in height; spicules in 8 rows; axial side concave, with a rudimentary tentacle; polyps bilaterally symmetrical.
L. gracilis, Verrill.	...	...	None.	Smooth.	...	Thin, with spicules.	At intervals of from 5–10 mm.; tentacles all of equal size, and short. Base often thicker than axis, and spicules longitudinally arranged.
L. fragilis, Wright and Studer.	Axis a creeping stolon.	...	None.	...	Axis feebly calcareous.	Very thin, almost without spicules.	1–5 mm. apart; 4–4·5 mm. in height; 8 rows of spicules on middle of polyp-body.

Chrysogorgia orientalis, Versluys. Plate VII. fig. 2.

This species is represented by two fragments, portions of a larger colony.

The axis of the main stem or branch is yellowish and iridescent, rigid and brittle, about 1 mm. in diameter. The branching is very profuse, and the arrangement is irregular.

One polyp is present on each node, but sometimes one occurs exactly at the point of divergence of two branches. The polyps are long and slender, 1·7 mm. in height and 1 mm. in oral diameter. They are broad at the base, then narrow, and again expanded. The spicules are arranged longitudinally, except around the base, where they are arranged obliquely transverse to the long axis of the polyp.

The spicules are spindles with small rough wart-like protuberances. They show marked variation in length, in the degree of roughness, and in shape, but all may be classed as spindles or flat sword-shaped spicules with serrated edges. Several show an X-shaped mark about the middle, as if they were incipient or reduced quadriradiate forms. Their measurements, length by breadth in millimetres, are as follows:

Spindles, 0·8 × 0·1; 0·7 × 0·1; 0·6 × 0·12; 0·5 × 0·12; 0·3 × 0·04; 0·25 × 0·05.

Flat, sword-shaped, 0·35 × 0·05; 0·23 × 0·05.

Locality: Station 2; 6° 32′ N., 79° 37′ E.; 675 fathoms.

Previously recorded from East Indian Archipelago (Ceram Sea and Timor Sea).

Chrysogorgia flexilis, Wright and Studer. Plate II. fig. 3.

This species, which is represented by several specimens, belongs to Versluys' Sub-*G*roup A 2. The colonies vary in height from 100–130 mm. They are bush-like or bottle-brush-like, and have a compact appearance. The basal attachment of one specimen is a thin, flat calcareous expansion, white in colour; in another case there were two root-like stolons.

The axis is brown with a greenish tinge in the lower part, and yellow with a marked golden lustre in the upper part of the stem and in the branches. In diameter it varies from 1–1·5 mm. at the base.

The branches are arranged in dextrorse 2/5 spirals. At the origin of each branch the stem is pushed a little to the opposite side and thus presents a zigzag appearance, which is more marked in the upper part of the stem than in the lower. It seems as if the stem were spirally twisted. The length of the internodes of the stem varies from 1·5 mm. in the lower part to 4·2 mm. in the middle and upper part, and the distance between two branches that lie directly over one another varies from 12 mm. in the lower part to 18 mm. in the middle and upper part of the stem.

The lower branches are broken off, but their origins are easily distinguished, except on a short basal portion about 7 mm. in length. Each branch is divided in dichotomous fashion with a more or less marked predominance of one of the components of the first division.

On the older parts there is one polyp to each internode, placed almost at the point of the next division, but sometimes a little below the bifurcation, especially in the younger parts of the colony. At the end of the twig there are two polyps, one at or near the tip and the other about half-way down the internode. The polyp has a large swelling at the base which almost surrounds the stem, it is then contracted, and it again swells out at the tentacular portion, thus appearing very like a globular vase with a neck. On the swollen portion of the polyp placed near the tip of the twig, the spicules are arranged transversely and then longitudinally, but in the others they are arranged longitudinally. On the aboral surface of the tentacles there is a band of spicules. The polyps vary in height from 1·9–2·3 mm.

The cœnenchyma is very thin and membranous, and allows the axis to shine through. Though much weathered, it shows in parts a single layer of closely apposed spicules

The polyp spicules are spindles, sharp or blunt at the ends, straight or curved, and covered by minute tubercles. The following measurements were taken of length and breadth in millimetres :

0·6×0·08 ; 0·5×0·06 ; 0·42×0·06 ; 0·25×0·05 ; 0·13×0·03 ; 0·06×0·02.

Some of these dimensions are large, for in the Siboga specimens they were 0·12–0·26–0·33 mm. in length by 0·06–0·07 mm. in breadth, and in the Challenger specimens 0·25–0·41 in length by 0·03–0·07 in breadth.

The polyp spicules include several irregularly shaped spindles, and a few incipient quadriradiate forms with a distinct X-shaped marking in the middle.

The spicules of the cœnenchyma are flattish spindles, blunt at the ends, and covered by small tubercles with a serrated appearance at the edges. A number have an X-shaped marking at the middle. Their measurements are as follows :

0·22×0·04 ; 0·14×0·03 ; 0×13×0·03 ; 0·06×0·03.

Embryos were found in several of the specimens.

Localities : Station 333 ; 6° 31′ N,. 79° 38′ 45″ E. ; 401 fathoms. Station 267 ; 7° 02′ 30″ N., 79° 36′ 10″ E. ; 457–589 fathoms. Station 254 ; 11° 16′ 30″ N., 92° 58′ E. ; 669 fathoms. Station 241 ; 10° 12′ N., 92° 20′ 30″ E. ; 606 fathoms.

Previously recorded from coast of Chiloe and East Indian Archipelago.

Chrysogorgia dichotoma, n. sp. Plate VI. fig. 3.

This species is represented by a damaged specimen, apparently the upper portion of a colony.

The close-set branches are arranged in a sinistrorse but quite irregular spiral.

The axis is very brittle and is slightly kneed at the origin of each branch, so that it has an apparently spiral course. It is brown with a tinge of green in the lower parts, and golden yellow in the younger parts of the stem and branches. It is very calcareous, and is cylindrical with a smooth surface.

The internodes are about 2 mm. in length, and the distance between two branches that stand directly over one another varies from 7–10 mm. Each branch divides dichotomously with a predominance of one of the elements of each bifurcation, and the branching of each main branch is almost wholly in one plane.

The polyps are very small, and arranged on the branches in short spirals. They are somewhat conical in shape, the larger 0·5 mm. in length by 0·25 mm. in breadth. The tentacles are long, and have an aboral band of spicules arranged longitudinally in two or three rows.

The polyp-spicules are spindles or rods, blunt or rounded at the ends, and covered by minute small spines. The following measurements were taken of length and breadth in millimetres:

0·16×0·02; 0·12×0·02; 0·11×0·015; 0·08×0·01.

The cœnenchyma is very thin, and shows no spicules. It allows the axis to shine through, throughout its whole length.

Locality: Station 237; 13° 17′ N., 93° 07′ E.; 90 fathoms.

Chrysogorgia irregularis, n. sp. Plate II. fig. 4; Plate IX. fig. 6.

This species is represented by a number of broken portions of a colony.

There are no polyps on the largest fragment, and the cœnenchyma is also almost entirely lost.

The axis is hard, brittle, and calcareous, iridescent and golden yellow, deep in tint in the lower part, but lighter in the upper part of stem and branches.

The branching partakes of the nature of a helicoid cyme, with the fourth branch often rising directly over the first; but there is no regular arrangement. There is usually one polyp for each node, but on the younger branches two are often found on a single node. Each polyp has eight projecting points, and is bell-shaped, the wide end closed by the infolded tentacles, which form an eight-rayed star, and have a coating of massive spicules on their aboral surface.

The spicules of the polyps are chiefly rod-shaped, and are arranged in a spiral on the body of the polyp. This arrangement gives place at the oral end to an arrangement by means of which eight points are formed from so many sets of converging spicules. The base of the polyp is considerably larger than the branch on which it stands, so that the branch seems to run through the base.

The spicules of the general cœnenchyma and of the polyps are very irregular in shape, some rod-shaped, some spindle-shaped, and some very irregular and flat. They are smooth, with the exception of a few of the more irregular forms, which show slight papillæ on their surface or at the edges.

Locality: Station 202; 7° 4′ 4″ N., 82° 2′ 45″ E.; 695 fathoms.

Chrysogorgia indica, n. sp. Plate III. fig. 6.

This species is represented by a number of very incomplete fragments, which present a striking appearance owing to the contrast between the white polyps and the deep bronze of the axis.

The axis is hard, brittle, and cylindrical.

The branching in the fragments is dichotomous and very profuse. The cylindrical polyps are of a beautiful white colour, and stand at right angles to the axis. The base forms an inverted V, which passes down the sides of the axis and almost meets at the other side. There is one polyp to each internode near the younger portions; in the terminal portions two may be present; in the older portions there may be two or three, and on one internode as many as four were seen. The polyp spicules are arranged longitudinally and regularly. On the aboral surface of the tentacles there is a band of longitudinally arranged spicules.

The cœnenchyma is very thin and membranous, and allows the axis to shine through in all its length.

The polyp spicules are spindles, straight or slightly curved, and rounded at the ends, or irregular flattish rods. They have small tubercles on the surface, and many of the larger forms seem smoother than the average, except that their edges present a marked serration. The following measurements were taken of length and breadth in millimetres:

0·8×0·2; 0·6×0·13; 0·3×0·05; 0·25×0·02; 0·2×0·05; 0·16×0·02; 0·1×0·04.

The spicules of the cœnenchyma are smaller than the polyp-spicules, but very similar in shape and appearance. The following measurements were taken:

0·35×0·05; 0·3×0·025; 0·3×0·05; 0·13×0·01; 0·12×0·02; 0·07×0·06; 0·06×0·02.

Locality: Station 334; 6° 57′ N., 79° 33′ E.; 568 fathoms.

Family *Isidæ*.

Subfamily Ceratoisidinæ.

Ceratoisis gracilis, n. sp.
Acanella rigida, Wright and Studer.
,, *robusta*, n. sp.

Ceratoisis gracilis, n. sp. Plate VI. figs. 6 and 6*a*.

This species is represented by several pieces, the longest of which measures 328 mm.

The axis is cylindrical, and consists of horny nodes and calcareous internodes. It is unbranched, long and slender, tapering gradually, and only slightly flexible in its lower part. The horny nodes are short, about 1 mm. in length, while the calcareous internodes are long, on an average 9 mm. in length in the largest specimen. The internodes are hollow, and are formed of concentric layers; their surface is smooth, and they are uniform in thickness except for a slight swelling just at their junction with the horny nodes. Through the centre of the horny nodes a slender calcareous rod runs, forming a junction between the adjacent internodes.

The cœnenchyma is thin and transparent, allowing the nodes and internodes to shine through. It contains minute spicules. The polyps are long and slender, and are arranged in a sinistrorse 2/3 spiral. On each internode there are polyps, and their apices are more or less directed towards one side. They have long supporting spicules on the walls, somewhat stronger on the abaxial side, and several always project beyond the retracted tentacles. The polyps are 3–4·5 mm. in length.

The tentacles are short, and have one row of short blunt pinnules on each side of the middle line, thus leaving a free space on the oral and aboral surfaces. The aboral free space is covered by a band of very minute flattish spicules, which are arranged with their long axes perpendicular to the length of the tentacle.

The spicules are of the following types:

1. Long spindles which support the polyps, 0·9–3·2 mm. in length, 0·07–0·1 mm. in breadth.
2. Short flattish spindles, 0·075–0·3 mm. in length, 0·02–0·03 mm. in breadth.
3. Short flat spicules with broad spathulate ends, 0·2–0·45 mm. in length, 0·04–0·08 mm. in maximum breadth.

This species is viviparous. Several embryos were found in one polyp. They are globular bodies, 0·6 mm. in diameter.

Locality: Andamans; 270–45 fathoms.

Acanella rigida, Wright and Studer. Plate IX. fig. 14.

A complete colony, bushy in shape, 103 mm. in height.

The branches arise from the horny nodes in verticels of two, three, or four. The first whorl of branches is given off at the fourth horny node, a distance of 23·5 mm. from the base.

The axis has calcareous root-like processes. The solid calcareous internodes are grooved, short near the base, varying in length from 3–6·5 mm., but long near the extremities of the axis and branches, often reaching in the latter a length of 20–21 mm.

The polyps are prominent, rigid, and arise singly. They are covered with fusiform spicules, and have a length of 3·3 mm. The tentacles are not capable of complete retraction.

Locality: Laccadive Sea; 703 fathoms.

Previously recorded from "Challenger" Stations 194 and 194 A, off Banda; 200 and 360 fathoms.

Acanella robusta, n. sp.

This species is represented by a large colony, 250 mm. in height, without any trace of basal attachment.

The colony is very bushy, the branches being given off from all sides of the stem. They arise from the main stem singly or in twos or fours, and from the primaries the secondaries arise either singly or in twos or threes. Anastomosis of the branches occurs in a few places, but it is by no means common.

The axis is composed of alternate horny and calcareous joints, the latter marked by longitudinal ridges few in number and often indistinct.

The polyps are irregularly disposed on the stem, and few in number. On the branches they are placed alternately on opposite sides, but in several cases they arise so nearly at the same level as to seem opposite.

The polyps are firm and rigid, 3–5 mm. in height, with a basal diameter of about 2 mm. A number of projecting points extend beyond the incurved tentacles. On the polyp body the spicules, some of which are visible to the naked eye, are arranged in two layers, the inner layer consisting of small spicules irregularly disposed, the superficial layer consisting of large spicules with no obvious arrangement, some transverse in the lower part, some longitudinal especially on the upper part, but altogether placed so as to form a firm protective and supporting covering to the calyx. In many cases they seem to be more numerous on the abaxial side of the polyp, but in the polyp near the end of a twig they seem to be equally developed on all sides.

There is a band of longitudinally disposed spicules on the aboral surface of each tentacle, each band consisting of 3–4 rows. On the lateral surfaces small spicules are arranged transversely, and closely packed together so as to form a complete coating.

The cœnenchyma is very thin, with a few small spicules scattered irregularly. Near the base of a polyp the spicules are somewhat larger.

The polyp spicules are straight or curved spindles, and a few smaller forms which may be called rods. They are all covered by small sharp tubercles. The following measurements were taken of length and breadth in millimetres:

3·2×0·3; 2·6×0·23; 1·7×0·1; 0·5×0·1; 0·2×0·03; 0·1×0·017; 0·04×0·01.

The smallest of these spicules come from the lateral surfaces of the tentacles; they are flattish, and have prominent tubercles on the edges.

The spicules of the cœnenchyma are rods, few in number and small in size, 0·14×0·025; 0·18×0·025.

This species differs from *Acanella rigida* in the arrangement of the polyps, and in the details of the spiculation.

Locality: Station 325; 18° 18′ N., 93° 25′ E.; 843 fathoms.

In another specimen the branches arise from the main stem in twos or in threes, and from the main branches the secondaries arise singly or in twos.

The axis is composed of alternate horny and calcareous joints, the latter many times longer than the former, and marked by longitudinal furrows. The first calcareous internode is very short, 3·5 mm., while the second measures 17 mm. in length. The lowest joint is calcareous, and from its base large root-like processes spread out, pure white in colour and very brittle.

The arrangement of the polyps cannot be definitely made out; in some places they seem to be placed more or less alternately, in other places they seem to be restricted to one side of the branch. They are long, 3·75 mm. in height, and stand up sharply from the axis. They are somewhat contracted near the middle, and then swell out again just below the base of the tentacles. The spicules are arranged on the polyps at an acute angle to the longitudinal axis, and none reach the whole length of the polyp. On the aboral surface of each tentacle there is a narrow band of two rows of spicules arranged longitudinally; on the lateral surfaces narrow bands formed of very small spicules run out to form supports for the bases of the pinnules.

The cœnenchyma is very thin, and contains minute scattered spicules.

The spicules of the polyps are straight or slightly curved spindles, and a few rod-shaped forms. All are covered by minute spine-like tubercles. The following measurements were taken of length and breadth in millimetres:

2×0·13; 1·4×0·12; 0·4×0·05; 0·22×0·05; 0·08×0·01.

Some of the larger spicules have a narrower part near the middle, and the smallest have more prominent spines. The spicules of the cœnenchyma are short rods, either blunt or pointed at the ends, and in some cases constricted near the middle. The following measurements were taken:

0·3×0·04; 0·25×0·045; 0·14×0·02.

Locality: Station 254; 11° 16′ 30″ N., 92° 58′ E.; 669 fathoms.

Family *Primnoidæ.*

Subfamily Primnoinæ.

Stachyodes allmani, Wright and Studer.

Stenella horrida, n. sp.

Thouarella moseleyi, Wright and Studer, var. *spicata*, n.

Caligorgia flabellum, Ehrenberg.

„ *indica*, n. sp.

„ *dubia*, n. sp.

Stachyodes allmani (Wright and Studer) = *Calypterinus allmani*, Wright and Studer. Plate II. figs. 1, 5*a*, and 5*b*.

Two fragments, evidently portions of a larger colony. The larger (*A*), which is branched, is 196 mm. in height; the smaller (*B*), which is unbranched, is 147 mm. in height.

The larger specimen differs from the typical habit of the species, inasmuch as the branches arise alternately on three sides of the axis.

The axis has at its lower end a diameter of 1·5 mm. in *A* and 1 mm. in *B*. It is very brittle, slightly more flexible towards the apices of the branches, and consists of a horny matrix filled with calcareous particles.

The cœnenchyma is so thin that the iridescent axis shines through.

The polyps are arranged in verticels of four, with a distance of 2 mm. between the origins of the verticels. This arrangement gives the polyps a somewhat bilateral appearance, although they arise from three sides only, leaving on one side a bare strip which is formed into a canal by means of the large flat polyp spicules.

The locality of this specimen confirms the suggestion made by Wright and Studer that *C. allmani* is a deep-sea species.

Locality: Laccadive Sea; 703 fathoms.

The collection includes a complete young colony belonging to this species. It is unbranched with the exception of a small stump near the base, which is the basal portion of a branch or twig, and it reaches a height of 41 mm.

The polyps are arranged in a bilateral manner, in verticels of 3–4, and the apex of the polyp when at rest is directed downwards.

The development of the covered channel-like groove on the free side of the axis is well seen. As in the large specimens, the cœnenchyma is thin and allows the axis to shine through. The basal attachment consists of a flat spreading portion.

Locality: Station 226; 8° 36′ 15″ N., 81° 20′ 30″ E.; 542 fathoms.

To this species we also refer a large number of damaged and broken fragments from the following localities:

Station 267 ; 7° 02′ 30″ N., 79° 36′ E. ; 457–589 fathoms. Station 334 ; 6° 57′ N., 79° 33′ E. ; 568 fathoms. Station 2 ; 6° 32′ N., 79° 37′ E. ; 675 fathoms.

Previously recorded from Reefs, Fiji.

Stenella horrida, n. sp. Plate V. fig. 13 ; Plate IX. fig. 3.

Two specimens of a beautiful colony. The larger, which lacks its base, is 53 mm. in height, with a maximum width of 38 mm.

The base of the smaller piece is a flat expansion, from which the stem rises. The stem is dark brown in colour, horny in texture, with a poor development of calcareous corpuscles. It has a diameter of 1·3 mm. at its lower end, and becomes lighter in colour and filiform towards the ends of the branches and twigs.

The branching is nearly in one plane, and is somewhat profuse. The polyps, which are arranged in an irregular close spiral all round the axis, vary in height from 1·5–2 mm., and have a basal diameter of 1·2 mm. The oral end is surmounted by 6–8 projecting spines.

The thin cœnenchyma shows flat scale-like spicules, and it does not allow the axis to shine through.

The polyp spicules are (1) flat scales; (2) spindles straight or curved ; (3) spindles sharply bent at an angle ; and (4) flat irregular spicules with one or two projecting spines. The following measurements were taken of length and breadth in millimetres :

(1.) 1×0·6 ; 0·9×0·6 ; 0·6×0·28 ; 0·5×0·3 ; 0·5×0·2.

(2.) 0·5×0·04 ; 0·36×0·04 ; 0·32×0·04.

(3.) 0·8×0·13 ; 0·8×0·1 ; 0·6×0·05 ; 0·6×0·04.

(4.) 1·5×0·4 ; 1·3×0·45 ; 1·2×0·5 ; 0·9×0·6 ; 0·85×0·4 ; 0·8×0·6.

Locality : Andaman Sea ; 112 fathoms.

[Comparative Table

COMPARATIVE TABLE OF SPECIES

Species.	Appearance of Colony.	Axis, branched or not.	Axis, Shape and Colour.	Axis, Texture and Flexibility.	Iridescence or not.
Stenella johnsoni, Wright and Studer.	...	Branched, irregularly dichotomous.	...	Older, dense, horny, and calcareous; younger, horny, and feebly calcareous.	...
S. gigantea, Wright and Studer.	...	...	Slightly curved and grooved.	Dense, brittle, calcareous.	Iridescent.
S. doederleini, Wright and Studer.	...	Irregularly branched.	Dark brown colour.	Hard, brittle.	Iridescent.
S. spinosa, Wright and Studer.	Bush shaped.	Irregularly and densely.	Deeply striated, dark brown.	Hard, brittle.	...
S. acanthina, Wright and Studer.	...	Irregularly in incomplete spirals.	Brown.	Hard, fibrous, with calcareous corpuscles.	...
Thouarella moseleyi, Wright and Studer.	Twigs in two directions.	Twigs may develop into branches.	Thin, somewhat flattened and yellow.	Slightly calcareous, flexible.	Surface shining.
T. hilgendorfi, Studer.	Twigs from three sides of stem.	Branched.	Oval, with parallel longitudinal furrows, brown.	Horny, calcareous, rigid, brittle.	Slight golden lustre.
Thouarella köllikeri, Wright and Studer.	...	Branched in three directions.	Oval.	Calcareous, brittle in lower part, more flexible above.	...

Polyp Shape.	Arrangement of Polyps.	Spicules of Polyp.	Pre-opercular Spicules.	Opercular Spicules.	Spicules of general Cœnenchyma.
...	In whorls of two.	Scales in three rows.	Four large, forming a frill.	Eight.	Scale-like, often imbricated.
...	In whorls of four.	Scales in four rows.	Four large scales.	Prominent, deeply keeled.	Scales irregularly quadrilateral.
Long, narrow.	In whorls of three or four.	Scales in four rows.	Eight scales with spathulate prominence.	Eight in number, project, folded back.	Semi-transparent scales, many with projecting central knob.
Long.	Irregularly, or in whorls of two or four.	Scales in four rows, plus one pre-opercular row.	Eight with hollow spathulate spines.	Eight in number, each folded on itself, not projecting beyond the pre-opercular.	In older, small overlapping; in younger, larger and imbricated.
In height 2 mm.	In whorls of three or four.	Scales in four rows, plus pre-opercular row.	Three large acutely spined	...	Oval discs.
Club-shaped, 1·5 mm.	On main stem, in an irregular row. On twigs at first, in spirals up to three, then opposite.	Four transverse rows, four longitudinal rows.	...	Short, spear-shaped.	Irregular four or five-sided calcareous scales which overlap. A lower layer of triangular or irregularly polygonal.
...	In spirals of three, but apparently opposite.	Six to eight transverse, five longitudinal rows.	Strong projecting spines.	Eight spear-shaped, forming a low cone.	Longish and oval, lower edge covered by upper edge of preceding.
Pear-shaped.	In short spirals.	Eight transverse rows, five dorsal and lateral, and two ventral.	With strongly projecting spines.	...	Irregular, three-cornered, or polygonal, or four-edged, or rounded, over-lapping; inner layer three cornered or polygonal.

COMPARATIVE TABLE OF SPECIES

Species.	Appearance of Colony.	Axis, branched or not.	Axis, Shape and Colour.	Axis, Texture and Flexibility.	Iridescent or not.
T. antarctica, Valenciennes.	Bottle-brush-shaped.	Branched.	Oval and twisted in a spiral, yellow.	Horny, calcareous, brittle.	Golden lustre.
T. affinis, Wright and Studer.	Brush-like.	...	Oval twisted in a long spiral, then a second spiral begins, yellow.	Hard, brittle, but flexible near the tip.	Iridescent and golden lustre.
T. variabilis (type), Wright and Studer.	...	Not branched.	Elongated oval, brownish-yellow.	Horny, calcified, firm, brittle below, flexible above.	...
T. variabilis, var. *brevispinosa*, Wright and Studer.	...	Twigs branched.	...	...	...
T. variabilis, var. *gracilis*, Wright and Studer.	Finely ramified.	...	...	Firm, brittle, and calcareous below, becoming horny and flexible above.	...
Thouarella brucei, Thomson and Ritchie.	Creamy-white colour; bushy.	Branches arise in *three* directions, twigs from all sides; irregular.	Cylindrical; tawny brown, with at places a yellowish sheen, honey-yellow in twigs.	Horny and calcareous, almost inflexible, but twigs very flexible.	Slight yellowish sheen.
T. moseleyi, Wright and Studer, var. *spicata*, n.	...	Branches arise on the two lateral faces, not strictly alternate.	Cylindrical, with *longitudinal ridges and grooves.*	Hard and tough, with a darker portion in the centre; only slightly flexible.	Iridescent.

OF STENELLA AND THOUARELLA—*continued.*

Polyp Shape.	Arrangement of Polyps.	Spicules of Polyps.	Pre-opercular Spicules.	Opercular Spicules.	Spicules of General Cœnenchyma.
Club-shaped, 2 mm. high.	...	Seven transverse, eight longitudinal rows.	Higher than broad, strong prominences, and median tubercle, which projects as short spine.	Triangular, pointed.	Irregular, triangular to four-cornered, with inequalities.
Pear-shaped, 2 mm. high.	At varying intervals, but never in whorls or opposite, usually in spirals of three or four.	Seven transverse, five dorsal and lateral, two to five ventral rows.	Nearly lancet-shaped, two rows with prominent middle tooth.	Eight lancet-shaped, concave exteriorly.	Upper layer, irregularly three or four-sided; lower layer irregularly polygonal.
Cup-shaped.	In short spirals of three.	Four to five transverse rows, three irregular dorso-lateral longitudinal, and two ventral rows.	This and preceding row with long spines which project far, from six to eight in number.	Eight fine, sharply bent, lancet-shaped.	One layer, thin calcareous scales, edges overlap.
Club-shaped, 2·5–3 mm. high.	Wide intervals, arrangement into spirals less evident.	Five longitudinal rows.	This and preceding row have short flat spines, eight in number.	...	...
Cup-shaped, 2 mm. high.	In close spirals of three.	Five transverse rows.	This and preceding row develop spines.	Small and high, triangular.	Upper layer, irregularly polygonal to four-cornered, with free edge sharply serrated. Lower layer, thin lamellæ four-cornered or irregular.
Pear-shaped, generally bent inwards to the axis.	Arise in all directions, and with no definite arrangement.	About five transverse and seven longitudinal rows.	...	About seven, ridged, usually with a narrow leaf-like wing.	...
...	On twigs in pairs *almost opposite.*	Flat, scale-like, in five transverse and four longitudinal rows.	*With eight strongly projecting spines which may be bifid.*	Form a *high* cone.	Flat, irregular, multi-tuberculate scales.

Thouarella moseleyi, Wright and Studer, var. **spicata,** n.
Plate III. figs. 2 and 4.

Several incomplete specimens, without basal attachment.

The axis is branched, calcareous in texture, iridescent, cylindrical, and marked by longitudinal ridges and grooves. It is very hard and tough, and on being broken across shows a darker portion in the centre. Its diameter varies from 1–1·5 mm. in the different pieces.

The twigs arise on the two lateral faces, although here and there a twig may be seen with its origin slightly approximated to one of the other surfaces. They usually curve more towards one face than to the other, and this is accentuated by the presence of a polychæte worm lodged on one side of the stem. They are arranged in a roughly alternate manner, but here and there two on one side alternate with one on the other.

Polyps arise singly and in close proximity on the lateral surfaces of the main stem and branches between the origins of the twigs; on the twigs they arise in pairs on opposite sides. They are stiff and rigid, rising at right angles to the axis, usually 1·5 mm. in height and from 0·6–0·8 mm. in maximum width. They are covered with flat scale-like spicules, in five transverse and four longitudinal rows. The pre-opercular row of spicules has eight projecting spines which are quite visible to the naked eye, and project considerably beyond the opercular surface of the polyp. The operculum forms a somewhat high cone.

The general cœnenchyma is thin, allowing the axis of the stem, branches, and twigs to shine through. It shows numerous flat scale-like spicules.

The spicules are flat, irregular multi-tuberculate scales, varying considerably in size and with slightly convex edges. The following measurements were taken of maximum and minimum diameter in millimetres:

$$0{\cdot}6\times0{\cdot}4;\ 0{\cdot}45\times0{\cdot}28;\ 0{\cdot}28\times0{\cdot}26;\ 0{\cdot}14\times0{\cdot}08.$$

Some of the spicules bear a long smooth projecting spine, and their dimensions are:

$$0{\cdot}7\times0{\cdot}3;\ 0{\cdot}6\times0{\cdot}35;\ 0{\cdot}5\times0{\cdot}3;\ 0{\cdot}4\times0{\cdot}25.$$

The tubercles or warts are arranged in a regular manner round the nucleus of the scale, which is always eccentric. The spine projects from a groove, and is very easily detached; it is sometimes bifid.

The following table of comparison with the typical *Thouarella moseleyi* summarises the chief points of difference, on account of which we establish a new and very distinct variety:

Thouarella moseleyi.	Var. *spicata*, n.
1. Twigs alternate.	1. Twigs not strictly alternate.
2. Polyps in short spirals of three, or placed opposite.	2. Polyps in twos placed almost opposite.

Thouarella moseleyi.	Var. *spicata*, n.
3. Operculum bluntly conical.	3. Operculum forms a high cone.
4. Axis smooth.	4. Axis markedly ridged.
5. Axis flexible.	5. Axis only slightly flexible.
6. Pre-opercular scales not spined.	6. Pre-opercular scales with strong spines, which may be bifid.

Locality: Laccadive Sea; 703 fathoms.

The typical *T. moseleyi* was recorded from off Kermadec Islands.

Caligorgia flabellum, Ehrenberg.

This well-known species is represented by several large specimens.

We need only note that there is considerable diversity in the closeness of the whorls of polyps to one another, and in the number of polyps in a whorl. We counted 4, 5, 6, and 8.

Locality: Station 333; 6° 37′ N.; 79° 38¾′ E.; 401 fathoms.

Previously recorded from near Mauritius, East Indian Archipelago, Japan, East Pacific off Central America.

Caligorgia indica, n. sp.

This species from the Andaman Islands is represented by a fragment of a colony, 112 mm. in length and 20 mm. in breadth. It is closely allied to *C. similis*, Versluys, and to *C. versluysi*, Thomson and Henderson.

The axis is iridescent, and marked by a number of longitudinal grooves and ridges. It has a diameter of 1 mm. at its lower end, and gradually dwindles until it becomes thread-like at the tip.

We are greatly indebted to Dr. J. Versluys for generously placing at our disposal the following notes on this species:

"The primnoid in Professor Alcock's collection differs from *Primnoa ellisii*, von Koch (= *Caligorgia verticillata*, Pallas[1]) in several noteworthy details, as shown in the following table:

Character of Alcock's Species.	Characters of *Caligorgia verticillata*, Pallas.
a. Dichotomously branched.	*a.* Pinnately branched.
b. On the thinner twigs the polyps are arranged in whorls of 2, less frequently 3.[2]	*b.* On the thinner twigs the polyps are arranged in whorls of 3, very seldom of 2; on the thicker branches the whorls mostly contain 4 polyps.

[1] For more particulars see my memoir on the *Primnoidæ* in the monographs of the Siboga Expedition. No. xiii. a (1906).—J. V.

[2] On the *thicker* branches whorls of three were most frequent.—J. A. T.

Character of Alcock's Species.	Characters of *Caligorgia verticillata*, Pallas.
c. On a centimetre of the twigs there are 8 to 9 whorls of polyps.	*c.* On a centimetre there are only 5 or 6 whorls.
d. The length of the contracted polyps is 0·75 mm. or less.	*d.* The length of the contracted polyps is 1 to 1·25 mm.
e. The distance between the successive whorls of polyps generally varies from 0·4 to 0·9 mm.	*e.* This distance is 1 to 1·5 mm.

Alcock's species is more delicate, with much smaller polyps; the whorls are more numerous. In the details of the polyps there are also some differences. In *C. verticillata* the exposed surface of the polyps is covered by four well-developed longitudinal rows of scales, a pair of abaxial and a pair of lateral rows. These last rows are somewhat reduced in Alcock's species, where they are each represented by only three scales, two distal and one basal scale. The inner lateral rows are much reduced in both species.

The new species may also be easily distinguished from all the dichotomously branched species of *Caligorgia* hitherto described, viz. *C. ventilabrum*, *C. modesta*, and *C. compressa*; it is more delicate, with smaller polyps, and none of these three species has so few polyps in each whorl, even on the thinnest twigs.

There are, however, two new species in the collection made by the Siboga Expedition under the direction of Professor Max Weber in the Malay Archipelago, which in their habit, the dimensions of the polyps, and the small number of polyps in each whorl, much resemble Alcock's species. These are described in my monograph on the "Primnoidæ of the Siboga Expedition"; in the present note I can only point out the more important differences between these species and this new species.

One of them, *Caligorgia minuta*, is distinguished by the much larger and less numerous scales in the polyps; the abaxial rows consist of only 5 scales each, against 7 in Alcock's species; and of the outer-lateral rows only one distal scale is left, against 3 scales in Alcock's specimen. The operculum in *C. minuta* is a much lower cone.

The other new species, *C. similis*, very much resembles Alcock's species; but still it seems to me, so far as I can judge from the small material at my disposal of both species, that they are different. The differences are stated in the following table:

Characters of Alcock's Species.	Characters of *C. similis*.
a. The polyps are arranged in whorls of 2, less often of 3.[1]	*a.* The polyps are arranged in whorls of 3, very rarely 2, on the thinner twigs; on the thicker branches the whorls number 4, perhaps also 5 polyps.

[1] On the larger branches *three* was the usual number.—J. A. T.

Characters of Alcock's Species.	Characters of *C. similis*.
b. On 1 centimetre length of a twig there are 8 or 9 whorls of polyps.	*b.* On 1 centimetre there are 7 or 8 whorls.
c. The length of the contracted polyps is 0·75 mm. or less.	*c.* The length of the contracted polyps is somewhat larger ; it varies generally between 0·75 and 0·81 mm., but sometimes rises to 1 mm.
d. The distance between two whorls varies generally between 0·4 and 0·9 mm.	*d.* This distance is somewhat less ; it varies between 0·3 and 0·5 mm.
e. In the polyps the abaxial rows consist of 7 scales, the outer-lateral rows of 3 scales only (2 distal scales and a basal, or perhaps 2 basals).	*e.* The same number of scales is found in the abaxial rows, but the outer-lateral rows are reduced to only one distal scale.
f. Of the scales of each abaxial row only two are extended over the sides of the polyp ; sometimes a third scale also shows a smaller lateral extension.	*f.* Of each abaxial row the 5 proximal scales are extended over the sides of the polyp and replace the missing outer-lateral scales.

The sum of these differences is, I think, of sufficient importance to warrant us in regarding Alcock's species as different from *Caligorgia similis.* The two species are, however, very closely allied. The scales in the polyps and the cœnenchyma are of the same type.

C. indica differs from *C. versluysi* in the smaller number of scales on the polyps. Each abaxial row is formed by 7 scales only, against 10 or 11 in *C. versluysi*, each outer lateral row by 3 scales against 5. The habit of the two is the same, and it is not impossible that *C. indica* is only a variety of *C. versluysi.*"

Locality : Andamans ; 270–45 fathoms.

Caligorgia dubia, n. sp.

A single fragment 7·9 cms. in height and 1·6 cm. in width.

The branches arise in one plane from the opposite sides of the chief axis. The distance between the origins of the branches varies considerably, from 6·5 to 13·5 cms.

The axis is slender (1 mm. in diameter at the base, thread-like at the tips of the branches), flexible, almost cylindrical, and marked by longitudinal ridges. It is horny in texture, shows concentric layers, and is abundantly impregnated with lime. The colour is yellowish, darker in the angles between the chief axis and the branches.

The general cœnenchyma of the stem and branches is thin and transparent, and shows a layer of flat scale-like spicules, longer than broad, arranged with their long axis parallel to the length of the branch.

The polyps are arranged in whorls of two or three, and may also occur singly between the whorls, but in no case are there more than three in a whorl. They are long and slightly club-shaped, and their apices are always turned in towards the stem or branch on which they stand. The polyp spicules are flat, irregular scales with many warty tubercles, and some of them have one of their edges toothed. They are arranged in nine transverse and four longitudinal rows. There are two rows on the abaxial surface and two, not so complete, on the lateral surfaces; the axial surface is free from spicules. The spicules of the first three transverse abaxial rows have spines on their free edges, and the same is true of those on the first two lateral rows on each side. The operculum forms a distinct sharp cone, and consists of eight slightly curved spear-shaped spicules.

The spicules of the general cœnenchyma, which form a complete covering for the stem and branches, are long narrow flattish scales with the nucleus eccentric. Their measurements, length and breadth in millimètres, are :

0·85 × 0·175 ; 0·8 × 0·1 ; 0·7 × 0·17 ; 0·6 × 0·2 ; 0·6 × 0·1 ; 0·25 × 0·07.

The polyp spicules are broader than the above and more like fish scales. Their measurements, maximum length and breadth in millimetres, are :

0·4 × 0·2 ; 0·35 × 0·2 ; 0·3 × 0·25 ; 0·275 × 0·175.

The spicules of the abaxial and lateral rows, which have spines on their free edges, have the following dimensions :—0·37 × 0·3 ; 0·33 × 0·3 ; 0·3 × 0·3. The spear-shaped opercular spicules average 0·6 mm. in length and 0·27 mm. in maximum width.

Locality : Station 333 ; 6° 31′ N., 79° 38′ 45″ E. 401 fathoms.

Family *Muriceidæ*.

Acanthogorgia aspera, Pourtalès (= *A. spinosa*, Hiles ?).
Paramuricea indica, n. sp.
Acanthomuricea ramosa, n. g. et sp.
,, *spicata*, n. sp.
Anthogorgia verrilli, n. sp.
Calicogorgia investigatoris, n. g. et sp.
,, *rubrotincta*, n. sp.
Placogorgia indica, n. sp.
,, *orientalis*, n. sp.
trogorgia rubra, n. sp.
¹ On t*mptogorgia bebrycoides*, von Koch.

Acamptogorgia bebrycoides, von Koch, var. *robusta*, n.
„ *circium*, n. sp.
Acis spinosa, n. sp.
Muricella bengalensis, n. sp.

Acanthogorgia aspera, Pourtalès (= **A. spinosa**, Hiles?). Plate II. fig. 2; Plate V. fig. 15.

This species is represented by two beautiful specimens, 185 mm. and 120 mm. in height.

The colony is branched in one plane, the branches being in some parts alternate, in other parts quite irregular. The polyps are densely crowded on the stem and branches, giving each branch a bottle-brush appearance. At the base there is a flat and thin disc of attachment. Very light brown is the general colour, but the polyps are almost white. In the larger specimen the axis has a basal diameter of over 1·5 mm., and is flattened in the plane of branching from a point a short distance above the base. In the smaller specimen one surface is flattened and marked by a broad shallow groove, while the opposite surface is rounded. This flattening is not so marked in the branches, and some of the terminal branches or twigs are almost cylindrical. The surface of the axis is marked by a large number of discontinuous longitudinal ridges.

The polyps are arranged all over the stem and branches; they are long, up to 6 mm. in height, with a basal diameter of about 1·5 mm. They are almost cylindrical, with a tendency to expand slightly near the oral end. They stand at right angles to the stem and branches, and are firm and rigid.

On the polyp body the spicules are arranged in eight rows, each showing sharp projecting points formed by the free ends of the spicules. The spicules of each row are arranged in pairs, each pair converging and forming an inverted V with the apex directed towards the mouth of the polyp. The spicules of each V are slightly curved outwards at the lower end, so that they pass over to, and interlace with, the basal portions of the spicules of the adjacent V on each side. Towards the apex of each row the spicules become slightly smaller, and then the row ends usually in three projecting overlapping spines, one markedly larger and stronger than the other two. Thus the apex of the polyp is surrounded by a number of points which project a considerable distance beyond the infolded tentacles.

The spicules on the aboral surface of the infolded tentacles form a flat eight-rayed operculum. On each tentacle they are arranged at first longitudinally in the middle line with spicules on the sides at right angles to those in the middle, then the median spicules fail and the side spicules form Vs which become gradually shallower until the spicules lie transversely and alternate with one another.

The cœnenchyma is thin, and is densely packed with spicules. The spicules of the polyps are slender spindles, straight or curved, and a few slightly flattened forms. All have spiny projections, which are very marked in some of the smallest.

The following measurements were taken of length and breadth in millimetres :

1. General polyp spicules, 1×0·1 ; 0·8 × 0·09 ; 0·6×0·05 ; 0·35×0·03.
2. Tentacular spicules, 0·18 × 0·04 ; 0·14 × 0·04 ; 0·12 × 0·04.
3. Those that project at the apex, 1·6×0·09 ; 1·45 × 0·1 ; 1·4 × 0.12 ; 1·3×0·1 ; 1·2×0·1.

The spicules of the stem are similar to those of the polyps except for their smaller size. Their measurements are :

0·8×0·095 ; 0·8×0·08 ; 0·5×0·06 ; 0·35×0·09 ; 0·3×0·05.

Among the spicules, both of the polyp and of the general cœnenchyma, there are some more irregular forms, with processes about 0·1 mm. in length, bearing small spines.

We do not think that there is any valid distinction between *A. aspera*, Pourtales, and *A. spinosa*, Hiles.

Locality : Station 333 ; 6° 31′ N., 79° 38′ 45″ E. ; 401 fathoms.

Previously recorded from Havana, and *A. spinosa* (Hiles) from Blanche Bay, New Britain.

[COMPARATIVE TABLE

COMPARATIVE TABLE OF

Name.	Branching.	Cœnenchyma.	Verrucæ.
A. spinosa, Hiles.	In two planes at right angles. One terminal polyp.	Thin and fairly smooth.	Cylindrical, elongated, somewhat expanded at the summit, placed perpendicularly to the stem, and thickly crowded on all sides.
A. flabellum, Hickson.	Fan-shaped in one plane, with much anastomosis.	Thin and rough, beset with tuberculate warts.	
A. inermis, Hedlund.	In one plane, anastomosis present.	Of medium thickness, finely granular under the lens.	Cylindrical, with scarcely visible projecting needles, eight bands on the body of calyx.
A. brevispina, Studer.	Few on one side, in one plane, flexible.	...	Spicules in eight rows each, consisting of two sets of converging spicules. Eight apical spines not projecting beyond the operculum.
A. breviflora, Whitelegge.	Few, alternate in one plane.	Extremely thin, axis visible through, few spicules longitudinally arranged.	Somewhat constricted at middle; with spicules arranged transversely at base, not distinctly seriate, peripherally at tip, surmounted at the apex by a series of long needle-like spicules.

SPECIES OF ACANTHOGORGIA.

Polyps.	Axis.	Spicules.	Colour.	Notes.
Long, with spicules in eight longitudinal rows, arranged *en chevron*, surrounded at summit by eight bundles of projecting spines, each bundle with two or three needles.	Horny, and brown in colour.	In cœnenchyma, mostly quadriradiate, but also longish spindles, with small spines; in polyps, spindles bent at the lower end, with small spines.	Dirty white.	Differs from *A. muricata* in having eight groups instead of eight single projecting spines.
Small, cylindrical, with slightly projecting spines at the free edge, crowded on most parts of the flabellum, and principally in plane of flabellum.	Horny, with a core of bead-like calcareous bodies in definite chambers; black in the older parts, brown in younger.	Firmly locked together and fused irregular tuberculate foliaceous clubs and branched spindles; in anthocodiæ slightly bent spindles, partly smooth, partly tuberculate, also dagger-shaped forms; also irregular foliaceous spicules both in the anthocodiæ and the cœnenchyma.	White.	Most nearly related to *A. inermis*.
...	Horny, black-brown; in twigs yellow brown.	Slightly bent warty or spiny spindles, clubs irregularly rounded or branched at the thick end.	Yellow brown.	Near *A. laxa*, Wright and Studer.
Cylindrical rising at right angles, arranged in spirals.	Yellow.	Long spiny.	Cœnenchyma and polyps black.	...
Alternately on the sides of the branches, at right angles.	Horny, rather brittle, blackish-brown, tips of branchlets yellowish-brown.	In cœnenchyma straight or curved spicules, with few blunt spines; in polyps the base and sides have curved spindles, with a few blunt spines at ends, and occasionally tubercles in middle; the deep seated forms are curved or bent, smooth, or nearly so, coronal with long free end smooth, and basal portion strongly tuberculate, simple and bent or bifurcated.	Yellowish-white.	"This species may be distinguished from other species of the genus by the small polyps and large spicules" (Whitelegge).

COMPARATIVE TABLE OF

Name.	Branching.	Cœnenchyma.	Verrucæ.
A. ceylonensis, Thomson and Henderson.	In one plane, both alternate and opposite.	Thin.	Spicules in eight rows *en chevron.*
A. media, Thomson and Henderson.	Irregular, but on the whole alternate, not in one plane.	Of medium thickness, with many projecting spicules.	
A. muricata, Verrill.	Flabellate, main stem divides not far from base into several principal branches, which diverge widely at first, and then ascend nearly vertically; these give off numerous lateral branches which diverge nearly at right angles, and repeat their own arrangement.	Thin, with a loose granular appearance, but no projecting spines.	Elongated, cylindrical, or somewhat expanded at the summit, with about eight very long, slender, sharp, projecting and diverging spicules, the projecting part being usually more than ⅓ the length of the calyx; eight longitudinal ridges frequently with projecting ends near the summit.
A. aspera, Pourtales.	Flabelliform, few widely divergent branches.	Thin, rough, with projecting points.	Elongated, slightly constricted towards the summit, which is considerably expanded, armed with eight bundles of projecting spicules, eight longitudinal rows, with projecting tips, mostly towards the base and on the ribs.
A. armata, Verrill.	Irregularly branched, somewhat in a plane, branches occasionally uniting.	Thin, filled with conspicuous white spicules.	Elongate, often curved, capitate or clavate, surmounted by eight groups of long divergent sharp spicules, with eight longitudinal ridges, with an irregular chevroned arrangement.
A. hirta, Pourtales.	Sub-flabellate, irregularly branching, branchlets flattened and expanded at tip.		Lobed, spinous, with spicules horizontal near the base, and then in eight vertical rows.

SPECIES OF ACANTHOGORGIA—*continued.*

Polyps.	Axis.	Spicules.	Colour.	Notes.
In twos, threes, or in loose spirals.	Horny.	Spindles and quadri-radiate forms.	...	...
Opposite or spirally, but younger polyps interpolated.	Horny, chambered, brownish-yellow in older, yellow in younger parts.	Spindles with few warts, tri- and quadri-radiate stars and golf-club forms, with long shaft free, and head with rough warts; also a few irregular sex- and quinque-radiate forms.	...	Similar to *A. ridleyi* in mode of branching and in the arrangement of the polyps, but like *A. muricata*, Hiles, in the disposition of the spicules on the calyces.
...	Yellowish-brown, strongly striated.	In cœnenchyma, small, rough, irregular, and rather large, very roughly warted, or spinulose fusiform spicules, frequently curved. In calyces, slender, elongated, warty fusiform spicules, either straight or curved.	Greyish-white.	Closely related to *A. armata*, but differs in having smaller calyces, with longer spicules, and in form of cœnenchyma spicules; and also differs from *A. aspera* in not being hispid.
Alternately on two sides of the stem and branches.	Brownish-yellow.	In cœnenchyma, acute fusiform spicules, often S-shaped, similar in polyps.	...	Studer doubts if Hedlund's *A. aspera* should be referred to this species.
...	Yellowish-brown.	In cœnenchyma, white rough curved fusiform spicules.	Ash-grey.	...
Distant on stem, more numerous on the branchlets, irregularly alternate, prominent.	...	...	Grey.	...

COMPARATIVE TABLE OF

Name.	Branching.	Cœnenchyma.	Verrucæ.
A. verrilli, Studer.	Sparse branches arise at right angles and then run parallel to the stem.	Thin.	4-8 mm. long, expanded at the summit, which is surrounded by eight bundles of projecting spicules.
A. longiflora, Wright and Studer.	Sparsely in one plane, short stem dividing into two principal branches. Apex occupied by a polyp.	Thin, transparent.	With eight longitudinal rows, each formed of two rows of converging spicules, from tips of which eight bundles of spicules project.
A. ridleyi, Wright and Studer.	Branches on three sides, of which two on opposite sides most developed. Tip of twigs occupied by a polyp and thickened.	Thin and transparent.	With eight longitudinal rows, cylindrical, expanded at end, eight projecting bundles, each bundle consisting of at most three spicules.
A. laxa, Wright and Studer.	Chiefly on two sides in irregular alternating series, forming angles of 70° to 90°.	Thick, with spicules in longitudinal series.	With eight longitudinal rows, in each two rows of needles converging, the angle of convergence becoming blunter near the tip, with a peripheral ring of projecting spicules. Cylindrical, expanded at end.
A. ramosissima, Wright and Studer.	Strongly and densely ramified, branches from three sides; in two opposite directions; in one plane they are stronger; angle of divergence 45° to 50°; bush-like.	Thin and transparent.	Not in eight longitudinal rows, eight bundles at mouth. Cylindrical, expanded at tip.

SPECIES OF ACANTHOGORGIA—*continued.*

Polyps.	Axis.	Spicules.	Colour.	Notes.
Arise in all directions, very numerous, especially towards the tips of the branches.	Slender, flexible, horny, brown.	In cœnenchyma curved spindles or spiny clubs. In verrucæ curved spindles.	Yellowish-white.	Near *A. armata*, Verrill, but distinguishable by the smaller verrucæ and the shape of the spicules. It differs from *A. muricata*, Verrill, in the much shorter spines projecting around the summit of the verrucæ.
At wide intervals on stem and branches in alternating series, at apices two to four crowded together; arising at right angles to the branches.	Fibrous, horny in branches, quite flexible, yellowish-brown.	In cœnenchyma, long slightly bent spindles and rods armed with small widely separated spines, placed longitudinally in twigs and branches, generally truncated at one end. In polyp, needle-like; in calyx wall, blunt at ends; tentacles with longitudinally placed spicules.	Yellowish.	...
In short irregular spirals of three; young polyps interpolated between the more mature ones.	Horny, fibrous, soft, flexible, and somewhat flattened in branches, brownish.	Cœnenchyma spicules placed longitudinally, curved spiny spindles, with lateral projections. In calyces, spiny spicules relatively thick.	Brown.	Like *Acanthogorgia* (*Blepharogorgia*), *schrammi* (Duch. et Mich.).
On opposite sides; a third row may occur, but one side always free; at distances of 2 to 2·5 mm.	Horny, fibrous, soft, flexible, flattened in plane of branching in stem.	In cœnenchyma, thick longitudinal series; the edges bear blunt teeth, may be slightly forked at an end. Polyps with peculiar forms constricted at the middle, and forming diverging angles at ends. Tentacle spicules thin, with sharp spines and prickles, thickened at one end.	...	...
On stem short spirals of threes; greatly crowded at tip.	Horny, fibrous, brown, in thinner parts yellowish.	In cœnenchyma, spindles with sharp spines longitudinally placed. In polyps, sharp spindles, at margin triradiate, club-like.	Yellowish.	...

COMPARATIVE TABLE OF

Name.	Branching.	Cœnenchyma.	Verrucæ.
A. truncata, Studer.	Branched in one plane; branches arise at 45°.	Rugose.	Cylindrical, 1–2 mm. in height and 1–1·5 mm. in diameter; spicules arranged in eight groups.
A. horrida, Studer.	Branched in one plane. Branches distant, from two sides, arising at right angles, then running parallel to the main stem.	Bristling with projecting points.	Short, cylindrical or somewhat constricted near the base. Beneath the projecting spicules around the oral opening there is a collar of transversely placed spicules.
A. muricata, Verrill, var. *indica*, Thomson and Henderson.	Very profuse, confined to one plane.	Very thin.	2–5·3 mm. in height, with a basal diameter of 0·9 mm., and a width of 1·1–1·2 mm. at the crown; there are eight projecting spicules at the free end.
A. australiensis, Hentschel.	Branched more or less in one plane; lateral twigs at right angles to main stem and larger branches, to which, however, the longer twigs soon become parallel.	Very thin, surrounding six flat longitudinal canals.	Cylindrical to cup-shaped; uppermost peripheral spicules of verrucæ are in eight groups, each with two projecting spines.
A. schrammi, D. and M.	Flabellate, reticulate, much branched.	Thin.	Cylindrical, slightly elongated, somewhat narrower at the base; with 5–10 very long projecting spines.

SPECIES OF ACANTHOGORGIA—*continued.*

Polyps.	Axis.	Spicules.	Colour.	Notes.
Occur in spirals around the branches.	Horny, flexible, yellowish-brown.	(*a*) Cœnenchyma—spindles sometimes club-shaped. (*b*) Verrucæ—bent spindles; those at the apex are long and bent at the base, covered with spines, which are often branched. (*c*) Tentacles—short spiny spindles arranged transversely.	Brown.	...
In short spirals over the whole cœnenchyma, so that the latter is almost hidden.	Thin, flexible.	(*a*) Verrucæ—spiny spindles feebly bent. (*b*) Cœnenchyma—(1) ramified, plate-like, with numerous sharp spines; (2) spiny spindles curved or forked.	...	The spicules of the cœnenchyma recall those of *Paramuricea.*
Densely crowded all round the stem and branches.	Dark brown, fading to light yellow in the younger parts; covered with close-set furrows at the base.	...	...	Closely resembles both *A. spinosa*, Hiles, and *A. muricata*, Verrill, but differing from the latter in polyps and spicules.
Irregularly disposed; crowded at the tips of the branches, more distant below; no true operculum, but spicules on base of tentacles.	Horny, brown or yellow grey, approaching white in twigs.	(1) Cœnenchyma—(*a*) simple, somewhat curved spindles, with few small warts 0·39 mm. in length; (*b*) spindles with strong lateral outgrowths leading to X- and H-shaped forms 0·11 mm. in length. (2) Verrucæ — simple spindles. (3) Tentacle base—small, curved, flat, irregular, 0·12 mm. in length.	White.	...
Alternate.	Very black.	...	White.	Near *A. ridleyi*, Wright and Studer.

COMPARATIVE TABLE OF

Name.	Branching.	Cœnenchyma.	Verrucæ.
A. johnsoni, Studer.	Dichotomous, in one plane, no anastomosis.	...	Bell-shaped, constricted at base.
A. grayi, Johnson.	Irregular, tendency to grow in one plane.		Sessile, cylindrical, very small; spicules arranged longitudinally.
A. hirsuta, Gray.	Branches nearly in the same plane.		Bell-shaped; spicules overlap in indefinite rows.

SPECIES OF ACANTHOGORGIA—*continued.*

Polyps.	Axis.	Spicules.	Colour.	Notes.
Scattered on both sides of branches.	Blackish-brown.	(*a*) Four branched basal rays in one plane, with a long spine projecting beyond the cœnenchyma (spine 0·018 mm. long, processes 0·012 mm. long). (*b*) simple bent spiny spindles 0·021 mm. long.	Pale violet.	...
Irregularly scattered. There is a definite operculum of eight pairs of spicules.	Pale brown, very slender.	(*a*) Cœnenchyma—fusiform, slightly bent. (*b*) Verrucæ — those at apex—basal portion geniculate, flattened and very rough, often branched.	Dark brown.	...
Distant. There is no definite operculum.	...	Similar to those of *A. grayi*, but the large spicules are not branched at the base.	...	...

Paramuricea indica, n. sp. Plate III. fig. 3; Plate IX. fig. 20.

This species is represented by a single specimen, which consists of two stalks rising from a tubular basal portion. The stalks are simple, except that one gives off a branch at right angles which terminates in a slight club-shaped ending formed by three diverging polyps.

The axis, dark brown on the outside, with a lighter core, is about 1 mm. in diameter at the base. It is purely horny, showing no trace of effervescence when treated with acid. The verrucæ arise from three surfaces of the stem, and are so arranged that the verrucæ of one row alternate with those of the other rows. They are bluntly conical in shape, and have their spicules irregularly arranged.

The polyps are almost completely retracted, but in several cases the tentacles form a dome-shaped projection above the mouth of the calyx. The tentacles are well protected by spicules arranged as a crown with points. There are three to four rows on the crown. The lower portion of the polyps contain ova and *embryos.*

The single branch and the two stems end in blunt points round which three divergently projecting polyps are arranged.

The surface of the whole specimen is rough, with projecting spicules.

The following measurements were taken of the length and breadth of the spicules in millimetres:

1. Spindles, either straight or curved, with prominent rough warts, 0·6 × 0·1; 0·5 × 0·12; 0·41 × 0·12; 0·35 × 0·08; 0·3 × 0·06.
2. Tri-, quadri-, and sex-radiate forms:
   Triradiate, 0·4 × 0·2; 0·3 × 0·2; 0·25 × 0·2.
   Quadriradiate, 0·4 × 0·15; 0·4 × 0·3; 0·35 × 0·17; 0·25 × 0·15.
   Sexradiate, 0·45 × 0·18; 0·3 × 0·2; 0·3 × 0·12.
3. Irregular forms, many of which may be regarded as spindles, with a foliar expansion on one side:

0·6 mm.	in length with foliar expansion length by breadth in mm.	0·12 × 0·12.
0·5 mm.	,, ,, ,, ,, ,,	0·16 × 0·08.
0·5 mm.		0·14 × 0·07.
0·3 mm.	,,	0·14 × 0·06.

This species is nearest *P. æquatorialis*, but the anthocodial armature is much less elaborate, and the spicules differ in detail.

Locality: Andamans; 265 fathoms.

Acanthomuricea **ramosa**, n. g. et sp. Plate V. figs. 1, 4, and 8 ; Plate IX. fig. 5.

This new genus is established for two interesting specimens in which the axis is horny, solid in the older parts, but hollow and very soft in the younger portions. The colony is branched in one plane, but small twigs may arise at right angles to the plane of branching. The branching is irregular, but often alternate. The cœnenchyma is thin, like rough bark, with imbricating scales. The verrucæ occur on all sides of the axis, either closely or at moderate intervals, usually leaving the dorsal surface free in the older portions of the stem and branches. The tentacular portion of the retracted polyp forms a prominent conical operculum. The verrucæ are prominent, upright, almost cylindrical, with the spicules arranged in rows (sometimes slightly irregular) ending in eight points. The spicules of the verrucæ are irregular plates with a projecting spine, irregular discs with divaricate edges, elongated forms with broad bases, and a few irregularly branched rods. Those of the general cœnenchyma are irregular discs, flat bars, and spindles with a large foliaceous expansion about the middle of one side.

This genus seems to be related to *Placogorgia*, but differs from it in the thinness of the cœnenchyma, in the size of the verrucæ, and in the tentacular operculum.

More detailed Description.

The genus is represented by a large greyish-black, upright, fan-shaped colony, 27 cms. in height, with a maximum width of 23 cms.

The base consists of a flat spreading stolon-like portion which surrounds a fragment of *Pleurocorallium*. Almost at its origin the main stem gives off a branch at right angles. The branching is in one plane, and for the most part roughly alternate, though this is broken through, either by two branches alternating with one, or by an approximation to the dichotomous type. At the tip of a branch there are two or three verrucæ.

The thin basal portion is remarkable in having single polyps scattered over its surface, and in giving off a small branch from the under surface which lies in the same plane as the main colony, but has its apex directed in the opposite direction.

The axis is horny, 5·75 mm. in diameter near the base. It is solid in the lower part, but soon becomes hollow. At first the central tube is of small diameter, but it gradually increases in size till in the younger branches and twigs the walls become so thin that a slight touch compresses the tube and puts the walls in contact with one another. The axis is easily broken, and has its surface, at least in the main stem and branches, marked by a number of longitudinal striations.

The verrucæ are prominent, almost cylindrical, at right angles to the surface, about 2·3 mm. in height, with a basal diameter of 1·5 mm. The wall of the verruca is composed of scale-like spicules arranged in rows and furnished with blunt projecting points. The tips of the rows form eight points encircling the base of the tentacular operculum.

The anthocodiæ are not capable of complete retraction; the tentacular part when in rest lies on the top of the verruca and forms a conical operculum. The spicules are arranged in a crown and points. The crown consists of a ring of transverse spindles three deep; each point forms a triangle, the sides of which are formed by two large spicules with the basal ends diverging in golf-club fashion and resting on a spicule shaped like an inverted U. In nearly all the points there is a third spicule in close connection with the sides of the triangle.

The general cœnenchyma is thin and practically composed of large irregular imbricating scales. In the younger parts of the branches the cœnenchyma spicules are somewhat longer and less scale-like.

The spicules of the anthocodiæ are spindles, straight or curved, with spiny projections. They vary in length from 0·2 to 0·8 mm., and in width from 0·03 to 0·06 mm.

The spicules of the verrucæ are irregularly shaped plates with a projecting spine, irregular discs whose surface is covered by papillæ and whose edges are divided into branched protuberances, elongated triangles with broad bases which are divided into several branched prominences, and a few irregularly branched rods. The following measurements were taken of length and breadth in millimetres:

1. Plates, 0·95 × 0·6; 0·8 × 0·5; 0·75 × 0·4; 0·7 × 0·3.
2. Discs, 0·7 × 0·6; 0·5 × 0·35; 0·5 × 0·3.
3. Elongated triangles, 0·7 × 0·5; 0·6 × 0·25; 0·5 × 0·3.
4. Rods, 0·6 × 0·17; 0·5 × 0·11.

The spicules of the general cœnenchyma are flat bars, irregular discs, and spindles with a large foliar expansion on one side. All are covered by rough warty papillæ. Their measurements, taken as above, are as follows:

1. Flat bars, 0·8 × 0·25; 0·7 × 0·2; 0·6 × 0·2; 0·4 × 0·1.
2. Irregular discs, 0·6 × 0·4; 0·4 × 0·26; 0·4 × 0·2; 0·3 × 0·25.
3. Spindles, 0·7 × 0·12 across the arm, and 0·3 across the foliar expansion.
   0·6 × 0·1 „ „ 0·25 „ „
   0·4 × 0·1 „ „ 0·2 „

Locality: Station 284; 7° 55′ N., 81° 47′ E.; 506 fathoms.

Acanthomuricea spicata, n. sp. Plate IX. fig. 12.

A small complete colony 55 mm. in height, greyish-white in colour.

The branching is mainly in one plane, but small twigs with two or three polyps arise at right angles to this plane chiefly from the main stem. The branches arise alternately at wide intervals. There is a flat spreading basal attachment.

The axis is horny, cylindrical, and marked by longitudinal striæ. Its basal diameter is 1 mm.

The verrucæ are crowded, and occur all over the stem and branches without any obvious arrangement. On one branch they leave a bare strip along one side, and bare patches are seen here and there on the stem. There are clusters of verrucæ at the origins of small branches, and there are always two divergent verrucæ at the tips. The polyps are long, 2·7 mm. in height even in the partially retracted condition. On the lofty conical anthocodiæ the spicules are arranged in a crown with long points. The crown consists of four to five rows of small transverse spicules; each point has three spicules, two converging and one in close connection. On the aboral surface of the tentacles there is a narrow band of longitudinally arranged spicules, and there are smaller forms closely packed on the sides and directed towards the bases of the pinnules.

The verrucæ are from 1·8 to 2·2 mm. in height and formed of broad spicules, the upper ends of which project slightly and produce a rugged surface.

The rough bark-like cœnenchyma is formed of one layer of rough scales—like ctenoid fish scales—which are dovetailed into one another.

The spicules of the verrucæ and of the anthocodiæ consist of plates with a spine, irregular discs, elongated triangles with broad bases, rods, spindles, and golf-club forms. All are covered by warty protuberances. The following measurements were taken of length and breadth in millimetres:

1. Plates, 1·0 × 0·45; 0·9 × 0·4.
2. Discs, 0·85 × 0·55; 0·7 × 0·35.
3. Triangles, 0·9 × 0·4; 0·8 × 0·3.
4. Rods, 0·6 × 0·1; 0·5 × 0·06; 0·3 × 0·05.
5. Spindles, 0·9 × 0·09; 0·4 × 0·06.
6. Golf-club forms, 1·1 × 0·12; 0·9 × 0·1.

The spicules of the cœnenchyma are spindles with large expansions (sometimes foliar) on one side, irregular discs, and somewhat flattened bars. Their measurements are:

1. Spindles, 1·0 × 0·2; 0·8 × 0·2.
   Foliar expansion of spindles, 0·4 × 0·12; 0·2 × 0·23.
2. Discs, 0·6 × 0·4; 0·6 × 0·35; 0·4 × 0·4.
3. Bars, 0·5 × 0·19; 0·45 × 0·175.

They are all covered by very rough warty protuberances.

Locality: Station 333; 6° 31′ N., 79° 38′ 45″ E.; 401 fathoms.

Anthogorgia verrilli, n. sp.

A complete whitish colony, 140 mm. high and 30 mm. broad, attached by a flat expansion to a small fragment of rock.

A slightly sinuous main stem gives off branches irregularly in one plane. The first branch arises at a distance of 70 mm. from the base, and a second alternating with it 5 mm. higher. For 30 mm. the stem is bare, and then come four branches, irregularly disposed, two on each side. After another 15 mm. of bare stem there are other two alternate branches. The branches arise at angles varying from 30°–90°. Some of the primary branches bear secondaries.

The axis is brown at the base, golden yellow in the branches. Its texture is very soft.

The verrucæ are few and scattered. A typical example is 2·25 mm. high, 1 mm. in diameter. It consists of a cylindrical body on which the spicules are irregularly disposed. The body is surmounted by an overlapping dome-like operculum, consisting of crown and points. The crown spicules are arranged in two (or three) transverse rows; the sharp points consist of eight pairs of spicules which converge so as to enclose very small apical angles, sometimes with a smaller spicule between them.

The polyps are completely retractile within the verrucæ, the tentacles being infolded. On the aboral surface of the tentacles the spicules are arranged in two converging longitudinal rows.

The cœnenchyma is fairly thick, with large transparent spicules irregularly arranged, and in some cases visible to the naked eye.

The spicules are straight or curved spindles densely covered with warts, blunt or tapering at the ends. The following measurements were taken of length and breadth in millimetres:

Cœnenchyma spicules, 1·8×0·175; 1·5×0·2; 1·3×0·15.

Tentacle spicules, 0·2×0·02; 0·175×0·015; 0·15×0·02.

Verrill's diagnosis of the genus *Anthogorgia*, so far as it is known to us, is somewhat vague. It reads as follows:

"Colony branched with slender elongated branches: polyp calyces strongly projected, of a tubular form, with an eight-rayed operculum consisting of a thin ectoderm in which long spindles are embedded at various angles. Cœnenchyma thin, with large warty spindles."

As it seems possible to include our specimen in the genus *Anthogorgia* we have done so, rather than add to the already long list.

As Studer's *Anthogorgia japonica* has only a *tentacular* operculum, we do not think that it should have been referred to Verrill's *genus*.

Locality: Andamans; 270–45 fathoms.

Calicogorgia investigatoris, n. g. et sp. Plate IX. fig. 10.

It is not without much hesitation that we have established this new genus. The specimens on which it is based are difficult to deal with, and they seem to belong to two species. In the form of the calyces and in the nature of the spicules they approach *Anthogorgia*, somewhat vaguely defined by Verrill; but in the arrangement of the spicules on the verrucæ and in the nature of the operculum they are quite distinct. In some respects they suggest *Muricella*, but they are separated by the nature of the spicules and the size of the polyps. We have therefore referred them to a new genus, *Calicogorgia*, probably related to *Anthogorgia*; the principal differences between the two genera are stated in the following table:

Anthogorgia, Verrill.	*Calicogorgia*, n. g.
Verrucæ arranged in spirals.	Verrucæ lateral, with occasionally one or two on the other surfaces.
Operculum composed of spicules embedded at various angles.	Opercular spicules distinctly arranged in a crown and points.
Spicules on the verrucæ irregularly arranged.	Spicules on the verrucæ arranged in eight bands, each of two rows.
Operculum may be inside the margin of the calyx.	Operculum forms a dome on the tip of the calyx.

The following diagnosis of *Calicogorgia* may be given:

Colony branched in one plane, with polyps arranged principally on the lateral surfaces of the branches. Branching irregular, but approximately alternate. Verrucæ prominent with spicules in eight bands and with a toothed margin. Operculum a prominent cone, consisting of a crown and points. Spicules of the cœnenchyma and of the verrucæ are warty spindles either straight or curved.

More detailed description of C. investigatoris.

A greyish-white colony, with a faint tinge of pink, 97 mm. in height and over 100 mm. in width. A specimen in the littoral collection is 235 mm. in height, 205 in width, with an axis 3·5 mm. in diameter near the base.

The colony is branched in one plane. At a distance of 14 mm. from the base, the main stem divides into two principal branches which give rise to the secondary branches and twigs, the secondaries arising in a roughly alternate manner. The verrucæ are arranged on the younger branches in a single linear series on the two lateral edges, but here and there, especially on the older branches, they are present on the other two surfaces. The tips of the twigs and branches are occupied by two divergent verrucæ with a small interval between them. There is a flat spreading basal attachment.

The axis is altogether horny, dark brown in colour, hollow in the younger parts. It is marked superficially by a peculiar network with pits in the meshes.

The polyps are long, 4 mm. in height, even in a somewhat retracted condition. Ova are abundant, and are arranged in a loose necklace on the mesenteric bands. The non-retractile portions of the polyps are wider than the apices of the verrucæ, and stand out like domes. The spicules on the anthocodiæ are arranged in a crown and points; the crown consists of 3–5 (usually 4) rows of slightly curved warty spindles; each point is triangular in form, and consists of one basal spicule and three pairs of converging spicules, of which the two outer pairs reach to about the same level. At the apex of the triangle there are some minute spicules.

On the aboral surface of the tentacles there is a band of longitudinally arranged small spindles, with a few arranged transversely.

The verrucæ are conical in shape with a broad base, and are slightly flattened in the plane of branching. The spicules are arranged in eight bands, each of which consists of two rows of converging spicules. The points of the bands project a little at the top of the verruca. The verrucæ vary in length from 1·9–3 mm., and have a basal diameter of slightly over 2 mm.

The general cœnenchyma is thin, and allows the dark brown axis to shine faintly through the coating of long spindles.

The spicules of the polyps and verrucæ are spindles, straight or slightly curved, covered by warty protuberances. Several of the smaller forms have sharper spine-like processes. The following measurements were taken of length and breadth in millimetres:

1·4×0·18; 1·2×0·22; 0·7×0·15; 0·4×0·03; 0·16×0·04.

The spicules of the cœnenchyma are shorter and thicker warty spindles. Their dimensions are:

0·9×0·16; 0·8×0·12; 0·26×0·06; 0·16×0·05.

Locality: Station 246; 11° 14′ 30″ N., 74° 57′ 15″ E.; 68–148 fathoms.

Calicogorgia rubrotincta, n. sp. Plate IX. fig. 9.

This species is represented by two complete reddish colonies, 139 and 96 mm. in height, 88 and 64 mm. in maximum breadth.

The main stem rises from a flattened basal portion. The axis is dark brown, cylindrical, with a diameter of fully 1 mm. It shows no effervescence when treated with acid, and is formed of layers of dark brown horny fibres with a core of a lighter colour. Towards the tip of the main stem and of the branches it becomes very flexible and thread-like.

The stem, after giving off a branch, divides into two principal portions, one of which predominates. The secondary branches arise from all sides of the main branch, thus giving a somewhat bush-like appearance to the colony.

The verrucæ are prominent blunt cones, and arise perpendicularly. They are arranged alternately on the lateral faces of the stem and branches; but here and there this arrangement is disturbed by the occurrence of a few on one of the other faces.

The spicules in the verrucæ project slightly. On the tentacles there are colourless spindles arranged longitudinally, and there is a basal collaret formed of one or two rows placed transversely.

The spicules of the general cœnenchyma are of two types.

1. There are large spindles, either straight or curved, covered with rough warts. They vary considerably in size, and may be simple or forked at one end. Their dimensions, length by breadth in millimetres, are:

$$3{\cdot}4\times0{\cdot}55;\quad 2\times0{\cdot}33;\quad 1{\cdot}22\times0{\cdot}3;\quad 0{\cdot}9\times0{\cdot}175.$$

They are dark red, light pink, or colourless.

2. There are smaller spindles, either straight or curved, with fewer and simpler warts. These are usually colourless, but may be pink or light red. They vary in length from 0·2–0·5 mm. and in breadth from 0·03–0·1 mm.

The two sets of spicules shade into one another, and intermediate stages are not infrequent.

Locality: Bay of Bengal; 88 fathoms.

Placogorgia indica, n. sp.

An incomplete specimen of a light brown colour, 120 mm. in height.

The branching is on the whole in one plane, but a stump is seen near the lower end of the stem which rises at right angles to the plane of branching. The branches are irregular, five rising from one side and three from the other. The larger branches give off smaller branches, repeating the same arrangement. The basal portion is wanting.

The axis is horny, dark brown in colour, with a lighter core which exhibits no sign of effervescence when treated with acid. It is somewhat oval in section, and has its surface marked by discontinuous longitudinal ridges. In the younger portions it is light yellow and soft.

The verrucæ are arranged on three sides of the stem and branches, leaving one surface entirely free throughout the whole length. They are short and truncate (about 0·5 mm. in height), and have a number of projecting points round the apex.

In the wall of the verruca the spicules are arranged longitudinally, and many have a broad end formed of two diverging processes.

In the anthocodiæ the spicules are arranged in a crown and points. The crown consists of four rows of spicules, the triangular points consist of two converging spicules, sometimes with one in between, with the apices directed towards the base of the tentacles.

The spicules of the polyps and verrucæ are (1) simple spindles, straight or curved; (2) spindles with a foliar expansion on one side; and (3) spindles or clubs with a branched foliar expansion at one end.

They are all covered by rough warty protuberances. The following measurements were taken of length and breadth in millimetres:

1. Spindles, 1·2×0·2; 1·0×0·2; 0·8×0·12; 0·4×0·09; 0·4×0·05.
2. Spindles with foliar expansion on the sides, 0·9×0·2, with foliar expansion 0·2×0·2; 0·7×0·13, with foliar expansion 0·2×0·12; 0·5×0·12, with foliar expansion 0·2×0·2.
3. Spindles or clubs, 1·0×0·3; 0·8×0·4; 0·5×0·3; 0·4×0·36 in maximum diameter.

The spicules of the stem are spindles either curved or straight, with many rough warts, and often divided at the ends. Their dimensions are:

1·1×0·15; 0·7×0·15; 0·5×0·1; 0·4×0·04.

Locality: Andamans; 270–45 fathoms.

Placogorgia orientalis, n. sp.

An incomplete specimen, 83 mm. in height. There is a slightly conical basal attachment, from which the main axis arises. The branches lie in one plane, and arise irregularly, in some parts alternate, in other parts not. The axis is horny, oval in section, and dark brown in colour, with a lighter core in the younger branches.

The verrucæ are arranged in a linear series, on each lateral edge of the stem and branches, those of the one series alternating with those of the other. There are always two at the tip of a twig. The anthocodiæ can be retracted within the verruca margin. The anthocodial spicules are arranged in a crown and points; the narrow crown consists of four transverse rows; each triangular point consists of two spicules diverging slightly towards their base and of one or two others alongside of them.

The verrucæ are short and conical, almost alternate, 1 mm. in height, slightly truncated at the tip. Their wall is formed of scale-like spicules, with those in the upper row slightly more spindle-shaped. They show considerable diversity as regards the size and shape of the spicules, but these are always similar to those of the general cœnenchyma at the same place. They can close in so as almost to hide the completely retracted anthocodiæ.

The general cœnenchyma is thin, and is practically composed of imbricating scale-like spicules. On some parts of the stem and branches they are more spindle-shaped. In most parts the axis shines through the cœnenchyma, but towards the tips of the branches this transparency diminishes and finally disappears.

The spicules are oval scales, spindles either thick or slender, irregular scales, very irregularly branched spindles, and club-shaped forms with much branched folia.

This species differs from *P. atlantica* in having more prominent and less crowded verrucæ, and also in the details of the spiculation.

Locality: Andamans; 270–45 fathoms.

Astrogorgia rubra, n. sp. Plate V. fig. 10.

A portion of a reddish colony, 65 mm. in height. It consists of a main axis, from which two branches arise. The larger branch arises at a distance of 30 mm. from the base, is 50 mm. in length, and bears a secondary branch.

The axis is horny, spirally twisted, chambered, and brownish in colour, with a white core.

The polyps are completely retractile within dome-shaped verrucæ. On the smaller branches the verrucæ are disposed in a regularly alternate manner; on the larger branches they are almost opposite, and the spiral twisting of the axis makes it appear as if the verrucæ occurred on all the four sides. The bare part of the stem on each side is occupied by a longitudinal groove, spirally twisted.

The verrucæ are prominent and dome-shaped when the polyps are completely retracted, but truncated cones when the polyps are partially retracted. The apical margin of the verruca is marked by eight points, and there are eight opercular parts, each composed of a few large spicules.

The cœnenchyma is fairly thick, and has a granular appearance due to the arrangement of the spicules. They are all of a pink colour, somewhat diverse in shape, and arranged in two layers. The inner layer is composed of (1) long thin warty spindles disposed longitudinally, and varying in length from 0·3–0·4 mm., and in width from 0·015–0·002 mm.; and (2) of short, thick, slightly warty blunt spindles, from 0·2 mm. to 0·25 mm. in length, and from 0·04 mm. to 0·05 mm. in width.

The outer layer is composed of (1) small warty clubs, on an average 0·08 mm. in length by 0·93 mm. in width; and (2) irregular double clubs placed perpendicularly to the surface, about 0·08 mm. in length by 0·04 mm. in width.

On the tentacles there are longitudinal rows of short, thin, slightly warty spindles, which vary in length from 0·09–0·1 mm., and in width from 0·008–0·01 mm.

Locality: Station 218; 6° 55′ 6″ N., 72° 55′ E.; 210 fathoms.

Acamptogorgia bebrycoides, von Koch. Plate VI. figs. 4 and 5.
[= *Muricea bebrycoides*, von Koch (1887), p. 52.]

The colony is branched in one plane, with frequent anastomosis. There is a slightly conical basal attachment, with a flat, spreading margin. The main stem divides into two principal branches at a point 4·5 mm. from the base, and these give off branches from the two opposite sides in a very irregular manner. Between the origins of the larger secondaries smaller twigs or branches arise which do not usually divide, but only bear polyps. The branching is continued till branches of the fifth order are reached.

The axis is horny, and somewhat oval in section.

The polyps are arranged irregularly, chiefly on three sides, most abundantly on the lateral surfaces. The infolded tentacles form a fairly high spiculate operculum which rests on the top of the verruca. The anthocodial spicules are arranged in a crown and points; the crown consists of 2–3 rows of transverse spicules; the points have two pairs of converging spicules at the base, and from the tip of these several run longitudinally upwards.

The verrucæ are high, almost cylindrical, slightly smaller in diameter at the tip. They are formed of spicules identical with those of the general cœnenchyma, and have a height of 1·75 mm.

The general cœnenchyma is fairly thick, and has a peculiar appearance, due to the projecting globular-like ends of the spicules.

The spicules are of three types: (1) warty spindles, either straight or curved; (2) irregular branched spindles, with large flat foliar expansions, somewhat resembling the "Blattkeulen" of *Plexaura*; and (3) a few irregular quadriradiate forms. The following measurements were taken of length and breadth in millimetres:

Spindles, 0·5×0·06; 0·4×0·06; 0·2×0·02.
Irregular forms, 0·25×0·2; 0·2×0·175; 0·12×0·11.

The foliar expansions are very irregular, appearing in some cases to be the expanded branches of the main spindle, in other cases consisting of several branches fused together. They may be nearly smooth at the edges, or produced into a number of spines. The last type presents a very peculiar appearance, with rough tubercles on the simple branches and stem, with warty protuberances and irregular ridges on the foliar expansions.

Locality: Station 246; 11° 14′ 13″ N., 74° 57′ 15″ E.; 68–148 fathoms.

Previously recorded from the Mediterranean and the Azores.

Acamptogorgia bebrycoides, von Koch, var. **robusta**, n. Plate III. figs. 7 and 8.

The colony is branched in one plane, most of the branches being given off at one side. The tips of the twigs are somewhat club-shaped, and occupied by three polyps. The general cœnenchyma is rough in appearance. There is no trace of a basal attachment.

The axis is horny, yellowish in colour, and marked by a number of longitudinal striæ. The centre is composed of chambers separated from one another by transverse walls, and containing a whitish substance which gives little or no trace of effervescence when treated with acid. The cavity is not quite central, but lies sometimes nearer one side, sometimes nearer the other.

The polyps occur all over the surface of the main stem or branch. The verrucæ are almost cylindrical, but somewhat squat, 1·5 mm. in height, with a basal diameter of 1·3–1·7 mm. A number of spines project round the apex.

The retracted anthocodiæ bear an almost hidden low conical eight-rayed operculum, which is formed by the infolded tentacles. The spicules on the anthocodiæ are arranged in a crown and points; the crown consists of 2–3 transverse rows; the points have their spicules *en chevron*, with a number of short, curved spicules at the base of each triangle.

The cœnenchyma is thick, and has an arenaceous appearance.

The spicules of the anthocodiæ are spindles, straight or curved, and covered by warty projections. Several are thickened at one end, and present a club-shaped appearance. The following measurements were taken of length and breadth in millimetres:

0·6×0·1; 0·45×0·04; 0·2×0·03; 0·15×0·02.

The spicules of the general cœnenchyma and of the verrucæ are (1) spindles, straight or curved, with warty projections, or with large, thorn-like projections covered by warts; (2) incipient quadriradiate forms; and (3) irregular forms, with two or three rays at one end. Their dimensions, taken as above, are:

1. Spindles, 0·6×0·046; 0·5×0·04; 0·45×0·04.
2. Quadriradiate forms, 0·4×0·25; 0·3×0·2.
3. Irregular forms, 0·4×0·2; 0·3×0·2.

The last kind of spicule is extremely irregular; some seem to be almost flat, with large processes extending from the edges; others seem to be plates, with spines projecting from them.

Locality: Station 333; 6° 31′ N., 79° 38′ 45″ E.; 401 fathoms.

COMPARATIVE TABLE OF

Name.	Branching.	Cœnenchyma.	Verrucæ.	Polyps.
Acamptogorgia fruticosa, Germanos.	Tree-like, in various planes, angles 40°–60°. Two terminal polyps.	Fairly thick, very rough, with spicular warts, large rough spicules, very coherent.	Cylindrical or wart-like, spirally arranged, very thickly distributed, sometimes in groups of three to five.	Small, quite retractile, mouth circular with a frieze of parallel spines sloping inwards.
A. acanthostoma, Germanos.	Few branches, tree-like, in one plane, stem and branches of almost the same thickness, cylindrical, branch angle 60°–85°, twig angle 58°–70°.	Thin.	Cylindrical or truncate, conical, spirally arranged, with eight spines projecting very much at the aperture.	White, quite retractile, with high conical operculum of sixteen spicules in pairs on the tentacles, each like a long boot; also small spindles, rods, etc., on the tentacles.
A. horrida, Hickson.	Not in one plane, nearly at right angles, apparently dichotomous.	Thin, and bristling with spines.	Not prominent, but surrounded by a crown of spines.	...
A. tuberculata, Hiles.	Few branches in one plane. Two terminal laterally placed polyps.	Thin, rough, "lumpy," owing to projecting foliar expansions of the spicules.	Low, cylindrical.	Closely placed on three sides of the stem and branches leaving the hack free; with low conical operculum.
A. spinosa, Hiles.	Branches arise at angles of from 60°–90°. Two opposite polyps at tips.	Fairly thick, and very rough.	Cylindrical.	Chiefly on the sides of the branches at intervals of 2 mm.; with low conical operculum.
A. arbuscula, Gray MS. (Wright and Studer).	Branched in one plane, branches thickened at the ends and forming an angle of 45°–50° with the stem.	Rather thick and rough.	In shape cylindrical or bluntly conical.	In spirals of three at intervals of 1–1·5 mm., but closely packed at apices; operculum forms a low cone.
A. alternans, Wright and Studer.	Slightly branched from two opposite sides, angle of branching 70°–80°, terminal branches thickened towards the apex.	Thick and rough.	Blunt, conical.	Sparsely scattered on the stem, more thickly on the branches, crowded at apices; operculum low and conical.

SPECIES OF ACAMPTOGORGIA.

Axis.	Spicules.	Colour.	Notes.
Horny, yellow brown to yellow, glisten-ing, longitudinal striæ, rigid to flexible.	Irregular warty plates or spindles with long vertical process, digitate in verrucæ and larger; the warts and tubercles are always granular.	Deep red.	With considerable resemblance to *Acamptogorgia* (*Muricea*) *bebrycoides*, von Koch.
Rigid, horny, yellow, with white core, quite translucent, chambered, rigid except in twigs.	A basal tuberculate piece or plate parallel to the surface, with a vertical tuberculate process; also small irregular spindles, some almost double; verruca spicules have a spindle-base transversely disposed and a vertical projecting shaft. The tubercles are never granular.	Whitish to grey.	Verrucæ very like teasel heads.
Horny, black or brown, with core divided into chambers by vertical partitions.	Tripods with rough tubercles on legs and smooth projecting spines; in calyces two limbs to each spicule, in cœnenchyma three, four also found; the limbs may be foliaceous and so interlock; also slightly bent tuberculate spindles.	...	...
Horny, brown in older parts, white at the apices.	Spindles slightly bent, with spiny warts along one side, and, on the convex side at one end or about the middle, complex foliar prominences. In collaret, curved spindles with few spines; in operculum, flat spiny spindles arranged in eight groups of three.	White.	Similar to *A. fruticosa*, Germanos, but polyps larger, and quite distinct in colour.
Horny, brown, with the central core divided into chambers.	In polyps, three-rayed with foliaceous expansions from two of the three rays; the third a spike standing perpendicularly to the others. In cœnenchyma, bent spindles with short branched expansions on the convex side, and also smaller forms of the polyp spicules.	Light brown.	...
Horny, brown, not longitudinally striated, slightly flattened on the stem, flexible in branches and soft.	In cœnenchyma, bent spiny papilliform spindles, frequently with short foliar expansions. In calyces, three-rayed stars; in collaret, roughened spindles truncated at one end.	Stem brown, oral region of polyps with violet tint.	...
Horny, feebly flexible, yellowish-brown.	In cœnenchyma, curved spicules, frequently with foliar prominences. In calyces, triradiate with third ray shortened and with a bifoliar appendage with dentate margins. In collaret, long-curved spicules with few spines. In operculum, flat, spiny, and armed at one end with smooth projections.	Whitish.	...

COMPARATIVE TABLE OF

Name.	Branching.	Cœnenchyma.	Verrucæ.	Polyps.
Acamptogorgia atra, Thomson and Henderson.	Confined to one plane; branches arise at right angles, but bend upwards and run roughly parallel to the main stem.	...	Stand perpendicularly at intervals of about 2 mm.; three or four occur at tips of branches, but none terminal.	Restricted to the lateral edges of stem and branches, but here and there on back and front.
A. bebrycoides, G. von Koch.	Branched in one plane, with frequent anastomosis.	Fairly thick; globular-like ends of spicules project.	High, almost cylindrical.	Chiefly on three sides, most abundant on lateral surfaces; operculum fairly high, spicules arranged in crown and points.
A bebrycoides, G. von Koch, var. *robusta*, n.	Branched in one plane; branches mostly arise from one side.	Rough; thick.	Almost cylindrical, but somewhat squat; a number of spines project around the apex.	Occur all over the surface; operculum low, conical, almost hidden.
A. circium, n. sp.	Irregular; branches arise from two or three sides; there may be abundant anastomosis.	Rough, thorny, bristling with projecting spines.	Prominent, 3 mm. in height, with a basal diameter of 2 mm.	Arise irregularly from all sides; operculum high, conical, crown consists of four rows, and there are two converging spicules in each point.
A. spinosa, Hiles, var. *ceylonensis*, Thomson and Henderson.	Profusely branched in one plane; anastomosis common.	Thin, and very rough in appearance under the lens.	Low, 0 8 mm. in height, with a basal diameter of 0 6 mm.	On three sides of the axis; operculum conical.
A. gracilis, Thomson.	...	Rough, with sharp projecting points.	Stand perpendicularly to a height of about 1 mm. At top there is a ring of spicules in two rows.	May occur on all sides, but are mostly lateral, sometimes opposite, sometimes alternate, and branch ends in a pair; opercular covering consists of eight pairs of curved foliaceous spindles.
A. rubra, Thomson.	Branched in one plane; twigs arise almost at right angles.	Thin, and very prickly.	Small, and arise at right angles to the axis.	Alternate or subopposite; two occur side by side at the end of a branch.

SPECIES OF ACAMPTOGORGIA.

Axis.	Spicules.	Colour.	Notes.
Horny, almost black at base, but light brown in younger branches.	...	Black.	...
Horny, oval in section.	(1) Warty spindles, straight or curved ; (2) irregular branched spindles, with large flat foliar expansions ; (3) irregular quadriradiate forms.	...	..
Horny, yellowish, marked by a number of longitudinal striæ ; chambered.	(1) Spindles ; (2) quadriradiate types ; (3) irregular forms.	...	..
Yellow, darker in older parts ; chambered by vertical partitions ; ridged —spirally on main stem and branches, longitudinally on secondary branches.	(*a*) Cœnenchyma ; (1) spindles straight or curved ; (2) curved spindles, with a foliaceous expansion about the middle of the convex side ; (3) U-shaped and irregular forms ; (*b*) polyps—spindles and clubs with divaricate folia, covered by spines which stand either perpendicularly or are inclined to the surface of the spicule.	...	Near *A. horrida*, Hickson, but differs from it in the size of the polyps and in the size of the spicules.
Oval in section in older parts, cylindrical in younger portions.	(1) Triradiate, with large irregular processes in the angles which often fuse, thus leaving plate-like portions with perforations ; (2) modified triradiate forms ; (3) spindles and club-shaped half-spindles, curved or straight, either with very rough projections or fairly smooth.	Dark near the base, but gradually becomes lighter in the younger parts.	...
Horny, brown at base, yellow in twigs ; chambered in some places.	(*a*) Curved warty spindles, with a bidentate or otherwise toothed foliaceous part from the middle of the curve ; (*b*) clubs with irregularly expanded divaricate ends ; (*c*) small irregularly stellate forms ; (*d*) forms with four or more rays.	Pinkish-red, with white papilla-like verrucæ.	...
Thin, yellowish.	(*a*) Straight warty spindles ; (*b*) narrow curved spindles, smooth terminally, warty about the middle ; (*c*) large triradiate forms, very irregular ; (*d*) smaller and often simpler triradiate forms ; (*e*) small, very warty spindles, with a foliaceous expansion about the middle.	Deep crimson.	...

Acamptogorgia circium, n. sp. Plate V. fig. 3; Plate IX. fig. 1.

Two specimens, 16 cms. and 20 cms. in length. There are flat, spreading basal attachments. The branching is very irregular. In one specimen the principal branches are confined to one plane, and the secondary branches are either in that plane, or arise at right angles to it. Of the secondary branches, those that arise on the track of the principal branches curl round, and come to lie either in the plane of the primary branches, or again at an angle to it, so that the branches on a whole are arranged in three directions. There is also abundant anastomosis. In the other specimen the branching is even more irregular, the secondary branches coming off from three sides; but there is no trace of anastomosis.

The horny axis is yellow in colour, darker in the older parts and lighter in the younger parts, and as far as examined shows no trace of lime. The core is divided into a series of chambers by vertical partitions, a structure usually well seen in this genus. The surface of the axis in the main stem and branches is marked by a large number of spirally arranged ridges; in the secondary branches the ridges run longitudinally.

The verrucæ are prominent, arising irregularly from all sides of the stem and branches, and each is surrounded by a crown of spines. They reach a height of 3 mm., and have a basal diameter of about 2 mm.

In the polyp the spicules are arranged in a ring of four rows round the base of the operculum, and the spicules in the high conical operculum are placed *en chevron* with two spicules in each point.

The surface of the verrucæ and general cœnenchyma bristles with projecting spines, so that the whole colony has a very rough thorny appearance, which suggested the specific name *circium.*

The spicules of the polyps are spindles and clubs with divaricate folia, covered by spines which either stand perpendicularly, or are inclined to the surface of the spicule. The following measurements were taken of length and breadth in millimetres:

1. Spindles straight or curved, 1·3×0·15; 0·9×0·075; 0·2×0·025.
2. Clubs with folia, 0·9–1 in length by 0·4–0·7 across the folia.

Many of the spindles have one end with spines arising perpendicularly to the surface, and the other end with slanting spines, especially in the curved forms. In the clubs, the length of the spine varies from 0·3–0·6 mm.

The spicules of the general cœnenchyma are (1) spindles straight or curved; (2) curved spindles with a foliaceous expansion about the middle of the convex side; (3) U-shaped spicules, and very irregular forms. The following dimensions were noted:

Spindles from 0·3–0·6 mm. in length, and from 0·02–0·05 in breadth.

U-shaped forms with a spread of 0·1–0·3 between the tips of the arms.

This species is near *Acamptogorgia horrida*, Hickson, but differs from it in the size of the polyps and in the size of the spicules.

The branches have much entangled foreign material, *e.g.* two specimens of a spirally coiled Solenogaster (*Rhopalomenia gorgonophila ?*).

Locality: Station 333; 6° 31′ N., 79° 38′ 45″ E.; 401 fathoms.

Acis spinosa, n. sp. Plate V. fig. 11.

A small specimen, white in colour with a tinge of brown, 86 mm. in height and about 50 mm. in width.

The branching is confined to one plane, and is very irregular. Between the origins of the large branches small twigs are given off bearing a few polyps.

The axis is horny, about 3 mm. in diameter at the lower end, slightly oval in section, with its surface marked by longitudinal ridges.

The polyps occur all round the stem and branches. In the anthocodiæ there is a very rudimentary collaret formed of at most two rows of spicules, and at the base of the tentacles two spicules are placed side by side, with their lower ends resting on the collaret, thus forming a conical operculum over the infolded tentacles.

The verrucæ are short, and the retracted anthocodiæ lie almost completely hidden within them. They are formed of scale-like spicules, and a number of apical spines project from the uppermost row.

The general cœnenchyma is thin and covered by large spicules, either spindles or elongated scales.

The spicules may be divided into three groups: (1) fusiform, (2) modified fusiform, and (3) squamous.

1. The fusiform spicules are usually straight, covered with rough warts, and have the following dimensions:—from 0·2–0·7 mm. in length, from 0·07–0·2 mm. in width.

2. The modified fusiform spicules are very large, and usually have a number of monticular processes on one side. They vary in length from 0·4–2·5 mm., and in breadth from 0·2–0·65 mm.

3. The squamous forms are very irregular in shape; they have several sharp-pointed processes growing out from one side, and are covered by rough warts. The following measurements were taken of length and breadth in millimetres:

1·1 × 0·6; 0·7 × 0·6; 0·5 × 0·4.

To the types of spicules already noticed there must be added:

1. Club-shaped spicules bent at one end, with the long handle nearly or quite smooth, 0·5–0·55 in length by 0·1–0·12 in breadth; and

2. Plate-like modifications of the squamous type, with an eccentric nucleus and with a large rough spine projecting from one side. Their dimensions are: 0·7 × 0·3; 0·7 × 0·4; 0·5 × 0·3.

Locality: Andamans; 270–45 fathoms.

Muricella bengalensis, n. sp. Plate I. fig. 8; Plate VIII. fig. 2.

A specimen of a rose-red colour, 33 mm. in height.

The colony is branched in one plane; the branches are almost regularly alternate. There is a flat-spreading base, and just above it the stem takes a half-S curve and then grows straight. The tips of the twigs are occupied by two divergent polyps. The axis is horny, with a faintly-striated surface, with a basal diameter of about 1 mm.

The polyps occur on the two lateral surfaces of the stem and branches, and are irregularly alternate. On the anthocodiæ the spicules are arranged at the base of the tentacles in inverted Vs, each formed of two converging spicules with their divergent bases resting on a transversely-placed spicule, which may have another lying close beside it. On the aboral surface of the tentacles there is a band of longitudinally arranged colourless spicules, 0·1 – 0·15 in length by 0·02 in breadth.

The verrucæ are short and truncated (about 1 mm. in height), built up of spicules arranged on the whole longitudinally. Round the apex there are projecting tips of spicules. The general cœnenchyma is thick, and contains long warty spindles either straight or curved, varying from transparency to pink and red. The following measurements were taken of length and breadth in millimetres:

2·8 × 0·3; 0·9 × 0·15; 0·36 × 0·07.

Locality: Andamans; 270–45 fathoms.

Another specimen is a brilliant red colony, 70 mm. in height by 45 mm. breadth, with a tendency to keep to one plane of branching. The polyps often appear alternate, but they are actually in a steep spiral.

The cœnenchyma is covered with large, rather broad, irregular spindles, covered with tuberculate warts. Some of them attain a length of over 2 mm. and a breadth of 0·4 mm.

The verrucæ are steep truncate cones formed by similar but smaller spindles arranged longitudinally. The apical margin of the verruca is irregularly interrupted by the projecting ends of the spicules forming the wall. There is a tentacular operculum formed of whitish spindles without much definiteness of arrangement.

Locality: Bay of Bengal; 88 fathoms.

Family *Gorgonidæ.*

Callistephanus koreni, Wright and Studer.

This species is represented by a portion of a colony, of a beautiful coral red colour, 83 mm. in height.

The main stem, or branch, is about 2·5 mm. in diameter, and gives off eight branches (6–28 mm. in length), two of which again branch. All the branches arise nearly at right angles, and lie for the most part in one plane. Two of the smaller branches and one of the twigs arise almost at right angles to the general plane of branching. The branching is more profuse than Wright and Studer supposed.

The verrucæ are arranged alternately on all sides of the main stem, but in the smaller branches they are more or less confined to the lateral faces. Wright and Studer describe them as altogether lateral. The termination of a branch or twig is occupied by two diverging verrucæ. The distance between two verrucæ on the same side varies from 3–4 mm. in the main stem, but is usually less in the smaller branches and twigs. The verrucæ have a maximum height and a maximum basal width of 2 mm. The polyps are capable of complete retraction; the tips of the tentacles are visible in a few.

The cœnenchyma is rough and of moderate thickness. The spicules composing it are warty spindles, club-like spicules, and a few irregularly quadriradiate forms. They vary in size as follows:

1. Warty spindles, 0·15–0·35 mm. in length and from 0·02–0·05 mm. in breadth.
2. Club-like spicules, 0·09–0·12 mm. in length, and from 0·04–0·07 mm. in breadth.
3. Quadriradiate forms, 0·13–0·16 mm. from tip to tip in one direction, 0·1–0·12 mm. in the other.

Locality: Andaman Sea; 238–290 fathoms.

Previously recorded from "Challenger" Station 344 off the Island of Ascension; 420 fathoms.

Family *Gorgonellidæ.*

Nicella flabellata (= *Verrucella flabellata*, Whitelegge).
Juncella elongata, Pallas.
,, *miniacea*, n. sp.
Scirpearella moniliforme, Wright and Studer.
,, *alba*, n. sp.

Nicella flabellata (= *Verrucella flabellata*, Whitelegge).

A small colony without basal attachment, ochreous-yellow to brownish-white in colour, about 95 mm. in height.

Most of the branches arise from one side of the main stem, which has a zig-zag course. The branches are slender, somewhat swollen at the tip, and confined to one plane. Most of the secondary branches are given off from one side of the primary branches.

The axis is horny and markedly calcareous, somewhat flattened in the plane of branching, with a shallow groove on the flattened faces. At the lower end it has a diameter of 2 mm.

The polyps occur in a sinuous row on each side of the younger branches; on the stem and on the older portions of branches they are more numerous and encroach on the two flattened surfaces, always leaving the slight median depression free. Those on opposite sides alternate.

The verrucæ are low conical warts, with the polyps completely retracted. In other species of *Nicella* the verrucæ stand out more prominently; but we disregard this difference, as we have seen in other Gorgonellids that the prominence of the verrucæ varies greatly according to the state of retraction and mode of preservation. The verrucæ are formed of spicules similar to those of the general cœnenchyma, and when the polyp is completely retracted they show no ray-shaped marking at the apex. They are about 1 mm. in height. The smaller branches, where the size of the verrucæ is large in proportion to the diameter of the branch, have an undulating contour.

The cœnenchyma is fairly thick, and has two layers of spicules. Next the axis the soft tissue of the cœnenchyma is very dark brown.

The spicules are small double clubs with warty projections and spindles. Some of the spindles are relatively broad, blunt at the ends, and closely covered with warts, others are very slender and sharp-pointed, and have relatively few spiny projections. A few showed a minute free space at the middle. The following measurements were taken of length and breadth in millimetres:

1. Small double clubs, 0·05×0·03; 0·04×0·03.
2. Spindles, 0·19×0·04; 0·14×0·03; 0·09×0·02.

The specimen is practically the same as Whitelegge's *Verrucella flabellata*, but it seems to us that this should have been placed in the genus *Nicella*.

Locality: Andamans; 270–45 fathoms.

Previously recorded from Funafuti.

Juncella elongata, Pallas. Plate I. fig. 10; Plate IX. fig. 17.

This species is represented by two branched pieces, 220 and 160 mm. in length. In both the base is present, and consists of a flattened portion, slightly conical in one specimen, spreading over the surface of a calcareous conglomerate.

The axis is calcareous, rigid, and brittle; it is slightly oval in section, but in the younger portions it becomes quite cylindrical and tapers till it is thread-like. It shows a very white core surrounded by a brownish cortex. It has a maximum diameter at the base of 3·3 mm.

The verrucæ are low and truncate, and arranged in four rows in the older portions. In the younger parts they appear to be arranged in two rows on the opposite lateral faces. They have a height of about 0·5 mm. and a diameter of 1·5 mm. at the base.

The cœnenchyma is thick, brick red in colour, and contains three types of spicules, of which the following measurements were taken of length and breadth in millimetres:

1. Spiny spindles, 0·13×0·025; 0·12×0·03; 0·1×0·03.
2. Double clubs, 0·09×0·05; 0·08×0·045; 0·065×0·04; 0·06×0·04.
3. Double stars, 0·08×0·04; 0·07×0·04; 0·07×0·35.

There is a trace of a bare space and a ridge-like mark; otherwise the specimens agree well with *Juncella elongata.*

Locality: Bay of Bengal; 88 fathoms.

Previously recorded from Atlantic (Pallas); West Indies (Ellis and Solander); variety from N.E. coast of Australia (Ridley); var. *capensis*, Algoa Bay (Hickson).

Juncella miniacea, n. sp. Plate V. figs. 7 and 12.

A fragment of a beautiful vermilion red colony, 43 mm. in height.

The stem, which lacks its basal portion, bears two large cirriped galls, one near the lower end and the other almost at the middle point. It gives off three alternating branches all in one plane. On the stem and branches the polyps are closely disposed and arise from all sides, even on the surface of the galls. The axis is very calcareous, straw-coloured, flexible, cylindrical, with a smooth surface and a diameter of 0·5 mm. near the lower end.

The verrucæ occur in a crowded irregular spiral, the fourth being often vertically above the first. They are large and prominent, ovoid in shape, and usually curved towards the axis. The polyps are wholly retracted, and in this condition the apex of the verrucæ shows a small median aperture marked by an eight-rayed star. They are 1·25 mm. in height, and have a basal diameter of 1 mm. and an apical diameter of 0·75 mm.

The general cœnenchyma is thin, lying evenly over the surface of the stem and branches; and also spreading over the barnacles, and giving rise to numerous small polyps there. The surface has to the naked eye a smooth but arenaceous appearance. Under low power it is seen to be continuously covered with the glistening spherical heads of slightly projecting radially disposed reddish spicules. The spicules include the following types:

1. Long warty colourless spindles, 0·175×0·03; 0·12×0·03.
2. Small warty reddish spindles, 0·08×0·02; 0·07×0·02.
3. Reddish single clubs, 0·055×0·03 (at the thick end).
4. Reddish double clubs, 0·08×0·04; 0·06×0·035.

This specimen belongs to the *Nicella* group of the Juncellas, but it differs from *Nicella* in the arrangement of the polyps. It agrees, however, in the form and termination of the verrucæ, and in the fact that the verrucæ arise at right angles to the stem and then curve inwards.

There are almost equal proportions of spindles and clubs.

Locality: Andamans; 120 fathoms.

Scirpearella moniliforme, Wright and Studer.

This species is represented by an unbranched fragment 88 mm. in length.

The axis is very calcareous and brittle, 2 mm. in diameter. There are two deep grooves on two opposite sides, and between them a number of smaller grooves.

The stem shows two marked furrows, caused by the two deep grooves on the axis, dividing the polyp-bearing portion into two narrow bands.

The polyps are arranged in four irregular rows, the members of which alternate.

The cœnenchyma is very thin. Its spicules are either warty spindles, spiny spindles, stars, or double clubs. The spicular measurements are nearer those of *S. gracilis*, but in the shape of the spicules and in the marked grooving of the axis and in the low verrucæ our specimen comes nearest *S. moniliforme*, differing, however, in having more than ten grooves on the axis.

Locality: 8 miles west of Interview Island, Andamans; 270–45 fathoms.

Previously recorded from Amboina.

Scirpearella alba, n. sp. Plate IX. fig. 15.

Three incomplete specimens, white in colour, 281, 411, and 408 mm. in length, with a corresponding diameter at the lower end of 1·75, 2·3, and 1·75 mm.

Two of the colonies are unbranched, but the longest branches at a distance of 251 mm. from the lower end.

The axis is cylindrical, hard, brittle, and very calcareous, but becomes flexible and filiform near the tips. It is marked by a number of grooves which run up for a short distance, and also by a number of small protuberances.

The stem is oval in section with a groove on the two flattened surfaces, faintly marked in two of the specimens.

The verrucæ occur in a single row on each side of the stem, those of one row alternating with those of the other. They are low and truncated (0·45 mm. in height), laterally compressed, with spreading bases. The diameter is 1·4 mm. at the base, 0·65 mm. at the apex.

The cœnenchyma is moderately thick, and contains spiny spindles, double clubs, and irregular stars with an X-shaped marking. The following measurements were taken of length and breadth in millimetres:

1. Spiny spindles, 0·19×0·07; 0·18×0·065; 0·15×0·04; 0·10×0·03.
2. Double clubs, 0·15×0·1; 0·14×0·08; 0·13×0·08.
3. Irregular stars, 0·11×0·08; 0·1×0·07; 0·1×0·065.

In the double clubs there are several grades, some with a fairly long bare middle part and few whorls of warts, others with a very short median bare part and many whorls of warts.

This species differs from *Scirpearella moniliforme* in the number of rows of verrucæ, in not having a deeply grooved axis, in the size of the polyps, and in the spicules.

Locality: Bay of Bengal; 88 fathoms.

ORDER V. STELECHOTOKEA, Bourne.

SECTION ASIPHONACEA.

Family *TELESTIDÆ*.

Telesto arthuri, Hickson and Hiles.
„ *rubra*, Hickson.

Telesto arthuri, Hickson and Hiles.

This species is represented by two fragments, one of which is attached by a broad basal membrane to a broken piece of coral. Both pieces are simple, unbranched, with numerous polyps. Their heights are 43 and 39 mm. They present a rough appearance owing to the projection of many of the spicules, especially at the oral surfaces of the polyps. The colour is white. The polyps are arranged in a very irregular manner, either in short spirals of 2–4, or in whorls. Between the spirals or whorls single polyps occur. A portion boiled with caustic potash disintegrates and leaves no axial skeleton. The following measurements were taken of the length and breadth of the spicules in millimetres:

1. Those with rough warts,
   a. Spindles straight or curved, 0·8×0·1; 0·6×0·15; 0·5×0·08; 0·3×0·05.
   b. Tri- and quadri-radiate forms, 0·6×0·35; 0·3×0·15.
2. Those with sharp spine-like projections,
   a. Spindles straight or curved, 0·8×0·07; 0·6×0·07; 0·5×0·04; 0·4×0·05.
   b. Quadriradiate forms, 0·2×0·18; 0·3×0·15; 0·4×0·16.

Many of the spicules seem to be a combination of 1 *a* and 2 *a*. They are bent at an angle, and have one of the ends covered with rough warts and the other covered with sharp spines.

The specimens agree with *T. arthuri* (1) in having no grooving, (2) in there being no axis, (3) in the arrangement of the polyps, and (4) in the shape of many of the spicules. They differ in not having the secondary polyps so crowded; but this probably means that the specimens are young forms, as is also indicated by the ready disintegration of the skeleton when boiled with caustic potash.

Locality: Station 232; 7° 17′ 30″ N., 76° 54′ 30″ E.; 430 fathoms.

Previously recorded from Blanche Bay, New Britain.

Telesto rubra, Hickson.

A small but complete specimen, 38 mm. in height. The base is formed by a flat spreading portion, which is attached to a piece of weathered rock.

There is a single branch given off at a distance of 34 mm. from the base.

The polyps, 2·5 mm. in height, arise at right angles to the stem, and are arranged on the four sides in such a way that they seem to form a spiral. Nearer the tip the polyps on opposite sides are almost at the same level.

After repeated treatment with boiling caustic potash the stem retains its form, and the polyp-walls remain quite intact.

The specimen agrees with *T. rubra* in the ridges on the stem, in the minute quantity of horny matter, and so on.

To this species we also refer another fragment 2·5 mm. in length. It is simple and unbranched, and has polyp calyces about 2 mm. in height.

Locality: Andamans; 120 fathoms.

This species has been previously recorded by Hickson from the Maldives: Mulaku Atoll, 25 fathoms; Mahlos Atoll, 23 fathoms; and by us from Trincomalee.

SECTION PENNATULACEA.

Family *PROTOCAULIDÆ.*

Protocaulon indicum, n. sp. Plate VII. figs. 3 and 7.

This species is represented by three complete specimens, 30, 44, and 46 mm. in height.

The stalk is about 14 mm. in length, and is very delicate. It has a spindle-shaped swelling just below the beginning of the rachis.

The rachis is long, and the polyps are placed opposite in a single row on each side. They are about 5 mm. in height; broad at the base, gradually narrowing, and then slightly enlarging again at the oral end. The tentacles are long and slender, with one row of pinnules on each side. Above their origin the oral surface of the polyp is swollen so that it projects a little. In the larger polyps the lower portion is filled with reproductive bodies.

The axis is thin, very calcareous, and quadrangular in section, each side showing a hollow groove. Near the lower end of the rachis in the largest specimen the axis is 0·25 mm. in diameter.

In the largest specimen the tip of the rachis is not occupied by a polyp, but just below the tip a polyp arises at right angles to the plane of the others. In a younger specimen two polyps occur below the tip.

Spicules are absent.

Locality: Station 239; 11° 49′ 30″ N., 92° 55′ E.; 55 fathoms.

There is only one other recorded species of this genus, viz. *Protocaulon molle*, Kölliker. The genus has a noteworthy geographical distribution from the "Challenger" Station 169, north-east of New Zealand, 37° 34′ S., 179° 22′ E., to the Andamans. Its bathymetrical range is no less interesting, varying from 55 fathoms in the present case to 700 fathoms for *Protocaulon molle*.

Family *Protoptilidæ*.

Protoptilum medium, n. sp.
Distichoptilum gracile, Verrill.

Protoptilum medium, n. sp. Plate III. fig. 1.

This species is represented by an incomplete specimen 133 mm. in length.

At the basal end there is a small globular swelling, the walls of which are very thin; above that there is a longer spindle-shaped swelling. At the upper end of the latter two grooves start, one running up the prorachidial surface, the other running up the opposite side. The prorachidial groove is more marked, and extends the whole length of the rachis; the other is not so evident, and gradually disappears.

The axis is cylindrical, smooth, brittle, and very calcareous. The superficial spicules of the cœnenchyma give the specimen a granular appearance.

At the lower end of the rachis the immature autozooids appear as one row on each lateral surface, those in the one row alternating with those in the other. As they reach maturity they gradually depart from the strictly lateral position, and become more dorso-lateral. The mature autozooids are closely appressed to the stem, 4·5–5·5 mm. in length from origin to apex, with an oral diameter of 2·5 mm. They are abruptly truncated at the upper end, but have no teeth. The tentacles have a few projecting spicules at their proximal end, and are capable of complete retraction, forming an eight-rayed star.

The spicules of the cœnenchyma are spindles or rods free from warts or spines, but ribbed. They are colourless and very translucent.

This species seems to be intermediate between *P. aberrans* (Kölliker) and *P. carpenteri* (Kölliker).

Locality: Station 151; Colombo Lighthouse, S. 64, E. 13½ m.; 142–400 fathoms.

Distichoptilum gracile, Verrill. Plate IV. fig. 7 ; Plate IX. fig. 2.

This species is represented by a large number of specimens, several of which have a length of 775 mm. ; but all are incomplete.

The axis is white in colour and very calcareous, varying greatly in diameter, from 0·7–1·5 mm. It is sub-cylindrical, with two opposite sides slightly flattened. Towards the lower end in several specimens two grooves are seen ; towards the upper end the shape is nearly quite cylindrical. It is very brittle, and only slightly flexible in the lower part.

The calices of the autozooids are prominent, but the axial side is closely appressed to and fused with the stem. In the lower part they seem to be arranged in spirals ; but this is soon lost, and they are arranged on opposite sides alternately. The abaxial edge of the calyx is toothed, but in some specimens this is indistinct owing to the state of retraction of the autozooids. The calyx is formed of small needle-shaped spicules arranged parallel to the long axis ; the length varies from 1·5 mm. on the lower part of the rachis to 3·6 mm. in the upper part. On the lower part of the rachis the calices are more crowded than on the upper part ; and the tip of the one reaches to the level of the base of the other ; but on the extreme upper portion of the rachis they are more widely separated.

In the majority of cases the autozooids are completely retracted within the calices.

The presence of siphonozooids in close connection with the polyps is an interesting feature. According to Verrill, three siphonozooids occur above each verruca ; Jungersen, on the other hand, from his study of the specimens in the Ingolf Expedition collections, was led to regard the presence of a third siphonozooid as quite problematical. From the specimens in the present collection it is evident that the siphonozooids may be two or three. There is one on each side of the calyx at about the same level, and over one of them there is sometimes a third.

The cœnenchyma is thin, allowing the axis to shine through in most of the specimens. Abundant ova are visible in the lower part of many of the autozooids.

The spicules are slender, smooth but fluted rods tapering slightly to both ends, but ending bluntly. The following measurements were taken of length and breadth in millimetres: 0·3×0·02; 0·3×0·04. There are smaller, more oval forms, 0·04×0·02. All the spicules are exceedingly brittle. In colour they vary from a very pale to a deep sherry tint. The upper portion of the abaxial wall of the calyx is reddish, all other parts are yellowish or yellowish-white.

In many of the specimens an *embryo* was found towards the base of the retracted autozooid on the abaxial side. The embryo is a flat circular disc from 0·52–0·55 mm. in diameter ; the ova are spherical, from 0·15–0·2 mm. in diameter.

Locality : Station 231 ; 7° 34′ 30″ N., 76° 08′ 23″ E. ; 836 fathoms.

A single specimen from another locality is more slender and graceful, and has a different colour scheme. It is whitish-yellow, with the tips of the calyces pinkish.

Locality: Station 321; 5° 4′ 8½″ N., 80° 22′ E.; 660 fathoms.

Previously recorded from: 63° 06′ N., 56° W.; 61° 39′ N., 17° 10′ W. (Jungersen); South-west of Nantucket Island; 39° 59′ 45″ N., 68° 54′ W. (Verrill); and 0° 4′ S., 90° 24′ 30″ W., 23° 59′ N., 108° 40′ W., 1° 7′ N., 80° 21′ W. (Studer).

The bathymetrical range varies from 700–1573 fathoms.

Family *Kophobelemnonidæ*.

Kophobelemnon burgeri, Herklots, var. *indica*, n.
Sclerobelemnon köllikeri, n. sp.
Bathyptilum indicum, n. sp.
Thesioides inermis, n. g. et sp.

Kophobelemnon burgeri, Herklots, var. **indica**, n.

This species is represented by a beautiful club-shaped colony, 57 mm. in height, with a pointed upper end.

The stalk is shorter than the club-shaped rachis.

The axis is white, almost cylindrical in shape, extending the whole length of the colony and tapering gradually to its lower end, where it has a diameter of about 0·5 mm.

The rachis is club-shaped, and tapers to a point at its upper end. It reaches its maximum diameter, 5 mm., at a point 4 mm. from its apex.

The autozooids are arranged irregularly on the pararachidial surfaces in five rows. They are of medium length, and are capable of complete retraction. At a certain stage of contraction they seem to have distinct calyces marked off from the upper part of the polyps. Their walls contain spicules, and in the lowest part of the walls these are very numerous, arranged longitudinally in eight bands with a few spicules between. Farther up these bands become narrow zigzag streaks with transversely placed spicules, but this transverse arrangement seems to be due to the contraction of the polyp walls. In the tentacles there are longer needle-shaped spicules arranged longitudinally.

The whole surface of the colony is covered by an outer coating of spicules which are visible to the naked eye and give a whitish appearance to the specimen. On the rachis this outer covering forms a honeycomb with the siphonozooids in the cells.

The siphonozooids are numerous, brown in colour, and present an eight-rayed appearance. They occur over the whole surface of the rachis not occupied by autozooids, with the exception of a narrow groove on the prorachidial surface, which extends the whole length of the rachis and has a markedly yellow colour.

In the autozooids the spicules are irregularly shaped rods blunt at the ends, often hour-glass-like. The following measurements were taken of length and breadth in millimetres:

Rods, 0·26×0·04; 0·08×0·07.

Hour-glass forms, 0·3×0·15; 0·2×0·1.

In the lower part of the stalk the spicules are flat, pear-shaped, spade-shaped, biscuit-shaped, three-cornered, and cruciform, the last two forms having a distinct X-shaped marking near the centre. In several of the biscuit-shaped spicules there is a distinct single mark running across the breadth, and in the pear- and spade-shaped forms the broad end is very often marked by small teeth. The surface of all the forms has a pitted or granular appearance. The following measurements were taken of length by breadth in millimetres:

0·17×0·05; 0·16×0·08; 0·15×0·055; 0·05×0·05.

This specimen differs from the typical *K. burgeri*, Herklots, in having the spicules arranged in bands on the autozooids, in the presence of markedly cruciform spicules, and in the larger size of the spicules.

Locality: Station 169; 13° 05′ 27″ N., 80° 33′ 44″ E.; 91 fathoms.

Herklots' type was obtained from Japan.

Sclerobelemnon köllikeri, n. sp. Plate VI. fig. 8.

If A. von Kölliker's separation of *Sclerobelemnon* from *Kophobelemnon* is justified, which we venture to doubt, this specimen should be referred to the former genus, for it has no spicules in its tentacles, and the autozooids are not arranged in longitudinal rows.

The single somewhat imperfect specimen is about 60 mm. in height, and is markedly club-shaped. The cœnenchyma of the stalk and of the lower part of the rachis is damaged, allowing the axis to project for 12 mm.

The axis is cylindrical, and tapers to the lower end. It is marked by a number of longitudinal furrows, two of which are considerably deeper than the others.

The rachis is longer than the stalk and considerably swollen near the tip, where it has a breadth of 10 mm. There is a prorachidial streak, 3 mm. in breadth, free from autozooids, densely covered by longitudinal rows of siphonozooids.

The autozooids are arranged bilaterally in about six short oblique rows, usually three in each row. The siphonozooids occur over the whole unoccupied surface of the rachis. As Kölliker describes in *S. schmeltzii*, there is marked contrast between the more delicate distal region of the autozooid and the more substantial calyx-like proximal region.

There is a very sparse occurrence of spicules, but some were found in the calyx-like lower portion of the autozooids. They are quite smooth and regular in outline, minute ovoid discs with a slight waist in the middle. Many suggest small double dinner-rolls. Those measured were 0·05 mm. in length and 0·03 mm. in breadth.

Locality: Station 246; 11° 14′ 30″ N., 74° 57′ 15″ E.; 68–148 fathoms.

Bathyptilum indicum, n. sp. Plate VII. fig. 4.

The specimen is broken into two pieces which make a total length of 118 mm. It consists of a rather long stalk and a shorter rachis, and is club-shaped apart from a large globular enlargement at the lower end of the stalk. The general colour is grey, with a pinkish tint more marked in the rachis and tentacles.

The stalk is long, and gradually increases in width towards the upper end. Its lower end bends at an angle of almost 90° with the main part, and bears terminally a large thin-walled globular enlargement. On its thin walls there are several white bands of minute white spicules.

The axis is sub-cylindrical, its flattened surface making it almost quadrilateral.

The rachis increases in size until near the tip, where it begins to taper. It ends, however, in a blunt point. The autozooids are arranged on the metarachidial and pararachidial surfaces, leaving the prorachidial surface with a broad free space.

There are five fully developed large autozooids and one or two smaller. They are large with long tentacles, and are slightly directed towards the tip of the rachis. They are 7 mm. in length, and the tentacles are slightly longer (8–9 mm.).

Small siphonozooids occur over all the surface of the rachis not occupied by autozooids. They project a little above the surface, and have their apices directed slightly towards the tip of the rachis.

At the base of the rachis there is a slight spindle-shaped enlargement containing reproductive bodies.

The general cœnenchyma of the stalk and rachis is thick, and has a granular appearance owing to the presence of abundant spicules.

The spicules are long slender fluted rods, with blunt ends which show a peculiar tooth-like arrangement when examined under a high power. The following measurements were taken of length and breadth in millimetres:

$0{\cdot}33 \times 0{\cdot}028$; $0{\cdot}28 \times 0{\cdot}02$; $0{\cdot}14 \times 0{\cdot}02$.

This species differs from *Bathyptilum carpenteri*, Kölliker, in the following points:

1. In having slightly smaller spicules, though the whole colony is larger;
2. In the arrangement of the siphonozooids;
3. In having a large bladder at the end of the stalk;
4. In having fewer autozooids;
5. In the size of the axis, which has a diameter of 1·05 mm.

Locality: Station 315; 10° 06′ N., 92° 29′ E.; 705 fathoms.

Thesioides inermis, n. g. et sp. Plate VI. figs. 1 and 2.

General Features.

This new type is marked by the greatly elongated slender rachis borne by a short stalk.

The autozooids are long, slender, and without calyces. They are not fused to form pinnules, but occur irregularly on the pararachidial surfaces.

The rachis is more or less quadrilateral, with grooves on the metarachidial and prorachidial surfaces, distinct on the lower part, less distinct in the upper part. There are no spicules.

This genus must be placed in the first section of the Junciformes of Kölliker, near the genus *Bathyptilum.* We have named it *Thesioides* from its resemblance to the raceme of *Thesium.*

More detailed Description.

The maximum length is 360 mm.

The short stalk shows a terminal enlargement in which the base of the axis lies, and another a short distance above this. The colour of the stalk is yellowish to pinkish-brown; the autozooids are brownish-white in their lower part, darker brown in their upper part and tentacles. There is a peculiar bluish tint all over.

The axis is quadrilateral; its sides show shallow grooves which become less distinct towards the upper end. The lower portion is reduced very rapidly, and lies curved into a hook in the lower swelling of the stalk; the upper portion tapers more gradually towards the tip, where it becomes filamentous.

The rachis is long and slender, more or less quadrilateral in cross section, and bears the autozooids irregularly on the pararachidial surfaces.

The autozooids vary in length from 2·5–10 mm. The longest have a diameter of 1·3–1·4 mm. They are probably cylindrical in shape, but they are greatly wrinkled owing to contraction in the preserving fluid. The tentacles are short, and bear on each lateral surface one row of pinnules, which, owing to crowding, seem to be arranged in two rows at the base. The tentacles themselves are very peculiar in shape, and may be divided into two parts; the lower part short and broad with the pinnules arranged as above, the upper part a little longer, thin and whip-like, with very few long flagella-like pinnules rising from it. On the lower part a number of the pinnules near the tip are also produced into long flagellum-like structures. This arrangement must have given the tentacles a peculiar appearance in the living state.

Siphonozooids are present on the pararachidial surfaces between the autozooids, and also form a row along each edge of the grooves on the prorachidial and metarachidial surfaces.

Spicules are entirely absent both in the stalk and in the rachis.

Localities: Station 324; 18° 0′ 15″ N., 93° 30′ 45″ E.; 448 fathoms. Station 323; 16° 25′ N., 93° 43′ 30″ E.; 463 fathoms.

Family *Umbellulidæ.*

Umbellula durissima, Kölliker.
,, *dura*, n. sp.
,, *intermedia*, n. sp.
,, *rosea*, n. sp.
,, *purpurea*, n. sp.
,, *elongata*, n. sp.
,, *köllikeri*, n. sp.
,, *radiata*, n. sp.
,, *pendula*, n. sp.
,, *indica*, n. sp.
sp.

Umbellula durissima, Kölliker.

A single specimen of this species was found at a depth of 1132 fathoms, near the Laccadives. Its total length was over 450 mm.

The long, cylindrical, flexible stalk is of a brownish colour, and apparently without siphonozooids. At the lower end there is a spindle-shaped enlargement, quadrangular in section, with prominent edges, 41·5 mm. in length. The axis is nearly cylindrical.

The rachis is a flat rhomboid expansion of the stalk, and is covered with siphonozooids, with the exception of a small narrow line on the prorachidial surface. They are very abundant at the bases of the autozooids.

The autozooids are four in number, three large and one smaller; their arrangement is by no means definite, but they seem to be arranged more on the meta- and para-rachidial surfaces. The average length of the body of the autozooid is 9·75 mm., that of the tentacles 11·3 mm. The bodies of the autozooids and the aboral surfaces of the tentacles are white, while the oral surfaces of the tentacles and the pinnules are brown. The white colour of the autozooids and the aboral surfaces of the tentacles is due to calcareous needles, which are very long and are easily visible to the unaided eye. They are arranged roughly in eight rows on the autozooid wall, continuous with the bands on the tentacles.

The spicules of the lower end of the stalk are small rods or ellipses, with longitudinally arranged ribs formed of minute spine-like processes. The following measurements were taken of length and breadth in millimetres:

$$0{\cdot}23\times0{\cdot}03;\ 0{\cdot}12\times0{\cdot}04;\ 0{\cdot}05\times0{\cdot}018.$$

The spicules of the autozooids may be divided into two groups, (1) spicules similar to those of the lower end of the stalk but longer and proportionally narrower and with fewer ribs; and (2) long spindles or rods with rough jagged ends and minute longitudinal ribs. The measurements, taken as above, are as follows:

1. 0·3×0·02; 0·2×0·02; 0·1×0·03.
2. 3·0×0·4; 2·2×0·23; 1·8×0·22; 1·7×0·3.

The present specimen agrees with the description given by Kölliker except in a few points, *e.g.*, some of the autozooid spicules are larger, and the arrangement of the spicules on the body wall is not so markedly in eight rows continuous with the tentacle rows.

The Indian Ocean specimen agrees closely with several collected by the "Scotia." The arrangement of the spicules in eight rows is somewhat disguised in the "Scotia" specimens, so that no weight can be attached to this point.

Locality: Laccadives; 1132 fathoms.

The geographical distribution of this species is very interesting, for it has been previously recorded from Station 234, North Pacific Ocean, south of Yeddo, Japan, by the "Challenger," and from 48° 6′ S. and 10° 5′ W. by the Scottish Antarctic ship "Scotia."

The bathymetric range is 565–1742 fathoms.

Umbellula dura, n. sp. Plate VIII. fig. 9.

The stalk is long (280 mm.) and flexible, with an enlargement at the lower end. It is quadrangular, and has a rough sandpaper-like appearance when dried. The axis is almost cylindrical.

The rachis is short, and forms an irregular inverted cone flattened laterally and gradually diminishing until it joins the stalk. The autozooids are few and large. In the smaller of the two specimens there are three, and as they appear to arise at the same level they may be in a circle; or, as the median one is smaller, it may be the terminal autozooid with two others lateral, and thus the arrangement may be bilateral. In the larger specimen there are four large and two very small autozooids, and they may be classed into three sets according to size, one very large, three scarcely so large, and of nearly equal size, and two small, again giving a type of bilateral arrangement, or an arrangement of one terminal and an imperfect whorl with two smaller polyps placed between the older.

The siphonozooids occur all over the rachis; they are small, and scarcely raised above the surface.

The spicules of the autozooids are of two distinct types:

1. Small rods blunt at the ends and covered by longitudinal ribs formed of pointed conical spines; and
2. Long rods either slightly pointed or blunt and slightly thicker at the

ends, with minute tubercles on the surface. The ends have a peculiar shattered or unfinished appearance.

The following measurements were taken of length and breadth in millimetres:

1. 0·14 × 0·04; 0·12 × 0·035.
2. 1·5 × 0·15; 0·95 × 0·125; 0·3 × 0·05.

The spicules of the cœnenchyma at the lower end of the stalk are short blunt rods or longish ovals. In addition there are several quadriradiate forms with a distinct X-shaped marking at the junction of the rays, as if the rays were dovetailed into one another. There are also a few four-cornered forms with an X-shaped mark. The surface of the rods is covered by small spine-like tubercles which are disposed in longitudinal rows and give the peculiar ribbed appearance. The measurements, taken as above, are as follows:

1. Rods, 0·14 × 0·04; 0·125 × 0·035; 0·115 × 0·04; 0·10 × 0·04.
2. Quadriradiate and four-cornered forms, 0·08 × 0·08; 0·06 × 0·05.

Locality: Station 315; 10° 06′ N., 92° 29′ E.; 705 fathoms.

Umbellula intermedia, n. sp.

The stalk is 220 mm. in length, slender and flexible, quadrangular in section, with a small enlargement at the very base and a long spindle-shaped enlargement a little above this. It has a maximum diameter of 1·2 mm., and is about 0·75 mm. in width at its narrowest part. The axis is quadrangular in section, with a groove on each side.

The rachis is distinctly bilateral and bears five autozooids, one terminal and two lateral on each side. It is ovoid in shape, and marked on the prorachidial surface by a ridge formed by the end of the axis which runs up to the terminal autozooid.

The autozooids are of medium length, with short bodies and long tentacles. They are rigid, and stand out stiffly from the rachis. The length of the body is 3·3 mm. and the tentacles have a length of 8·5 mm. On the body of the autozooid and the aboral surface of the tentacles there is a coating of spicules visible to the naked eye.

The siphonozooids occur all over the rachis, with the exception of the ridge on the prorachidial surface. They are small, and occur in small meshes in the spicular covering.

The cœnenchyma of the stalk, the superficial layer of the rachis, and the surfaces of the autozooids and tentacles are densely packed with whitish spicules visible to the unaided eye.

The spicules of the rachis and autozooids are rough rods or bars, blunt at the ends and ribbed. They vary from 0·15–0·35 mm. in length and from 0·015–0·03 mm. in width.

The spicules of the stem are smaller than those of the autozooids and proportionally wider. They are flat, and have a number of ribs running longitudinally. The following measurements were taken of length and breadth in millimetres :

0·14×0·05 ; 0·16×0·045 ; 0·09×0·05.

This species is intermediate between *Umbellula güntheri*, Kölliker, and *Umbellula leptocaulis*, Kölliker.

Locality : Station 278 ; 6° 52′ N., 81° 11′ E. ; 1912 fathoms.

Umbellula rosea, n. sp. Plate V. fig. 5.

This species is represented by a single greyish-white specimen, which had a beautiful pink colour when alive. The total length was 185 mm.

The stalk is very long and flexible, quadrangular in section, marked by a groove on each surface, and never exceeding 1 mm. in diameter. The axis is quadrilateral, and so deeply grooved on the sides that the wing-like angles seem to be joined by a small central portion. On the stalk there is a terminal enlargement and another longer one slightly above.

The rachis is very short, distinctly bilateral, with two pairs of lateral autozooids and one terminal. The autozooids have a length of 5–6 mm. even in the slightly contracted state, and a diameter of 2·5 mm. at the base. The tentacles may reach a length of 6 mm.

Siphonozooids are present on the metarachidial surface, and extend between the bases of the autozooids on to the pararachidial surfaces, but are apparently absent on the prorachidial surface.

In the lower part of the stalk there are abundant short and thick ellipses with numerous ribs. The following measurements were taken of length and breadth in millimetres :

0·16 × 0·03 ; 0·10 × 0·04 ; 0·09 × 0·03.

In the autozooids a few spicules can be seen in the skin ; they are longer and proportionally thinner than those of the stalk, rough at the ends and at the margin, the rough appearance being apparently due to the frayed edges of the ribs. Their measurements, taken as above, are as follows :

0·3 × 0·25 ; 0·26 × 0·03 ; 0·22 × 0·02.

Locality : Station 117 ; 11° 58′ N., 88° 52′ 17″ E. ; 1748 fathoms.

Umbellula purpurea, n. sp. Plate VIII. fig. 3.

The specimens representing this species were dredged from various depths in the vicinity of the Andamans.

The stalk is long, flexible, quadrangular in section with prominent edges, so that the sides of the stalk seem to form shallow grooves. It has a yellow colour. At the lower end there is a distinct enlargement with very prominent angles, and from this point the stalk tapers until a short distance below the rachis, where it again forms a distinct enlargement which gradually increases in size till it joins the lower end of the rachis.

The axis is quadrangular in section, with the sides so deeply grooved that the whole appears like four united pillars. At its lower end the axis is 2 mm. in diameter; it gradually tapers to the tip, which is hidden in the rachis.

The rachis is short, and has the appearance of a swollen part of the stalk. It carries a large number of autozooids, which all reach the same height or nearly so, the lower ones being the longest. The rachis is bilateral, although the crowded state of the autozooids almost entirely hides this arrangement. One specimen from which they were cut off showed that they stood in four irregular circles or whorls, the outermost containing 7, the next 12, the third 14, the fourth 10, and in addition there were 7 autozooids at the very tip that had no definite arrangement. The autozooids are very long, 42 mm. in maximum length, with a diameter of 3 mm. The tentacles are long and tapering, with one row of bluntly conical pinnules on each side. In the basal portion of the autozooids there is an abundant supply of ova about 0·75 mm. in diameter. The autozooids are yellowish-white to transparent in the lower parts, and greenish-black in the upper part; the aboral surface of the tentacles has a bluish tinge, and the oral surface and pinnules is light brown.

Siphonozooids are numerous on the basal part of the rachis; they are prolonged into Λ-shaped points on the pararachidial and metarachidial surfaces, and they completely cover the prorachidial surface.

Spicules are absent throughout the specimen.

Locality: Andamans.

Umbellula elongata, n. sp. Plate VII. fig. 6.

This species is represented by one complete colony, about 631 mm. in length.

The axis is more or less quadrangular in the lower part, and looks as if it were formed from four pillars fused where they touch. It becomes circular or oval in section in the upper part. It has a diameter of 2 mm. near the base.

At the lower end of the stalk there is a large swelling, which reaches its maximum diameter, 10 mm., at a point 86 mm. from the base. From this point upwards the stalk, which is thickly covered by siphonozooids, gradually tapers till it reaches the beginning of the rachis. On the lower part the cœnenchyma is thick, but on the upper part it is thin, except at the two opposite sides, where it forms a continuous fold which runs the whole length of the axis.

The rachis is long, gradually thickening from its junction with the stalk and then continuing without any increase in size. It is radial in symmetry, and carries a large number of autozooids which arise from all sides. The lower part is corkscrew-shaped, leaving only the upper part straight. It reaches a length of 145 mm.

The autozooids are apparently arranged in short spirals. They are of medium length, 18 mm., with a diameter of 2·5 mm. near the base. The walls are thick and tough, but in the majority the tentacles are missing; they are white in colour, with a slight violet tinge in parts. Ova and *embryos* are present in most of them. (See Plate VII. fig. 6.)

Siphonozooids occur all over the rachis, and throughout the whole length of the stalk. On the lower part of the stalk they are larger than those on the upper part and than those on the rachis. Those on the lower part vary from 0·3–0·5 mm. in height, and those on the rachis have a height of 0·28 mm.

Spicules seem to be restricted to the lower part of the stalk, where there are small rods smooth and rounded at the ends, and also a few four-cornered forms. The length and breadth of the latter is about 0·007 mm.

Locality: Station 229; 9° 29′ 34″ N., 75° 38′ E.; 360 fathoms.

Umbellula köllikeri, n. sp.

The stalk is quadrangular, in some parts oval in section. At the base of the stalk there is a small swelling about a millimetre in thickness. The stalk is about the same thickness throughout its whole length, but it has a very slight swelling at its junction with the rachis. It is continued up the ventral surface of the rachis, and forms a prominent spine below the place where the terminal autozooid is situated. Its length is about 39·5 mm.

The autozooids, five in number, are arranged bilaterally, but owing to the twisting in the rachis they seem at first sight to be irregular. The two lowest are larger than the others, and rise from the edges of the ventral surface, so that there is a little of the ventral surface exposed between them, while the two smaller appear to rise more from the dorsal surface, and have their bases closely approximated. On the ventral surface the siphonozooids are seen on the small space exposed between the bases of the autozooids and on either side of the axis. On the dorsal surface they are arranged about the bases of the autozooids and on the sides of the axis, the middle line being apparently free. Where the siphonozooids occur the spicules seem to have arranged themselves so as to form a slight circular ridge around them. The length of the rachis is 11·5 mm., of an autozooid with expanded tentacles 12·5 mm., of the body of an autozooid nearly 5·5 mm.

There do not appear to be any siphonozooids on the stalk.

The spicules of the autozooids are somewhat triangular in section, with one of the sides flatter than the others, and with a granular appearance throughout the whole length.

Locality: Station 118; 12° 20′ N., 85° 8′ E.; 1803 fathoms.

Umbellula radiata, n. sp.

This species is represented by a complete colony, 857 mm. in length.

The stalk is long, and has a large swelling at the lower end, which extends for a distance of about 213 mm. upwards. From this point the stalk gradually tapers till it reaches the beginning of the rachis, where it seems to run up the centre, and does not become evident on either of the surfaces. On the lower swelling, which has a diameter of 7 mm., four prominent ribs are present, two of which can be traced throughout the whole length of the stalk. The stalk at the upper end of the swollen portion has a diameter of only 1·75 mm.

The axis is roughly quadrangular, with the angles rounded off. On one of the sides in the lower portion a groove is present; in the upper portion it becomes almost cylindrical.

The rachis is longer than usual, and bears a large number of autozooids, which are arranged in whorls. The lowest whorl has four.

The autozooids are long and slender (46×2·5 mm.), with moderately long tentacles; they are brownish in colour, with the upper portion almost black.

The tentacles are long and tapered, with one row of pinnules on each side. The pinnules are short, rather blunt, and widely separated.

The surface of the autozooids and the tentacles is covered by a brownish layer, which is honeycombed in appearance, and easily rubbed off. On treating a portion of this with acid, no effervescence was observed.

The siphonozooids occupy the whole surface of the rachis between the autozooids; they are small, brownish, wart-like projections. On the swelling at the lower end of the stalk there are several smaller projections, which may be siphonozooids.

Among the typical autozooids there are two or three polyp-like bodies, 20 mm. in length by 4·5 mm. in width. Tentacles are represented by small blunt conical projections, in one case with seven minute pinnules present on each side. In these polyps the stomodæum and gastric filaments are short and simple, whereas in the typical autozooids they are long and convoluted.

No spicules were found either in the polyps, the rachis, or the stem.

Locality: Andamans; 490 fathoms.

Umbellula pendula, n. sp.

This species is represented by one complete colony, which has a total length of 744 mm.

The stalk is very long, and has a large swelling at the basal end, which extends for a considerable distance, nearly 170 mm. upwards. From this point, however, the stalk tapers gradually towards its upper end.

The axis is somewhat oval in section, and brownish in colour. Near the lower end it has a diameter of 1·9 mm., but it tapers abruptly towards the base, where it ends in a short coil, and is thread-like. At the upper end it becomes cylindrical, and gradually tapers. It has a diameter of 0·75 mm. at a point just below the beginning of the rachis. In the lower part it is rigid, in the upper part it is quite flexible.

The pendulous rachis is 55 mm. in length, and stretches on the concave side of the bent stalk as a thin membrane. In its lower part the prorachidial surface is marked by a stiff ridge, formed by the prolongation of the axis; thus it is somewhat bilateral in shape; but this bilateral symmetry is soon lost, and a radial arrangement assumed.

The numerous long autozooids are arranged in irregular whorls. They may attain a length of 40 mm., and have a basal width of 2·5 mm. There are long slender tentacles, with one row of pinnules on each side. The autozooids are creamy white in their lower parts; the dark bluish-black stomodæum shines through the upper part.

Siphonozooids are abundant, occurring all over the rachis, with the exception of the short ridge on the prorachidial surface of the lower part. They are small (0·2–0·3 mm. in height), and appear as minute conical projections on the surface.

Spicules are entirely absent, none being found either in the polyps, the rachis, or the stalk.

Locality: Andamans; 188–220 fathoms.

Umbellula indica, n. sp.

This species is represented by a complete specimen, measuring 706 mm. in length.

The stalk tapers gradually from a little above its base to its junction with the rachis. On the lower part there is a large swelling, which extends for a considerable distance up the stem. It has four wing-like projections, and is hollow. The lower part of the swelling is smooth, but the upper part is roughened by the siphonozooids, which project like warts above the surface. The siphonozooids can be traced almost up to the junction with the rachis.

The axis is quadrangular, with the sides slightly convex, and measures 2 mm.

in diameter near the base. From this point it tapers very quickly towards the basal end, and becomes quite thread-like and hook-shaped. Towards the upper end it tapers very gradually, and becomes almost completely cylindrical near its junction with the portion contained in the rachis.

On the lower part of the axis the cœnenchyma is fairly thick, and is developed into four wing-like folds; on the upper portion it is very thin, and closely attached to the axis, but the wing-like folds can be traced almost to the beginning of the rachis.

The rachis seems to be radial in symmetry, its lower end forming a slender inverted conical portion, on which no autozooids, but numerous siphonozooids are found.

The autozooids are long and slender, with the lower part packed with ova. They are light in colour, and contrast with the dark bluish-black colour of the upper portion, which is almost certainly due to the decomposed contents of the stomodæum. They are as much as 48 mm. in length by 3 mm. in width. The stomodæum and gastric filaments usually extend for about 14 mm. In the upper portion the stomodæum is slightly convoluted, but lower down it is merely wrinkled. The tentacles are long, slender, and tapering, with one row of delicate pointed pinnules on each side.

The siphonozooids cover the whole surface of the rachis not occupied by the autozooids. They are small projecting wart-like bodies, and in the parts not so directly subjected to the action of the preserving fluid are slightly brownish in colour. They are on an average 3·2 mm. in height, and are slightly oval in shape.

No spicules are present in polyps, rachis, or stalk.

Locality: Andamans; 238–290 fathoms.

Umbellula, sp. ?

The stalk is cylindrical, thin, 355 mm. in length. There is a distinct swelling at the lower end, quadrangular in shape, with the edges prolonged into fine wing-like portions. The length of this enlargement is 68 mm. Its greatest breadth from the edges of the wings is about 2·5 mm., but tapers gradually at both ends, more gradually at the upper than at the lower end.

The stalk gradually tapers from the top of the enlargement, and varies in thickness from 1 mm. to 0·8 mm. Towards its upper end it gradually thickens and forms a plate-like structure continuous with the rachis and flattened laterally, thicker on the one surface than on the other, so that a cross section would appear somewhat wedge-shaped. This portion of the stalk, together with the rachis, has a hook shape in two specimens, but in the other two it is slightly curved towards the dorsal side. About half-way up the stalk there is a peculiar bend or knee in all the specimens. The stalk is devoid of siphonozooids except at the junction with the rachis. It is yellowish in appearance, and has a distinct sarcosome.

The rachis is thin, and wedge-shaped, with a ridge. The autozooids appear to be irregularly arranged in the older forms, but in the younger they are bilateral. The rachis is covered with wart-like siphonozooids beneath the part bearing autozooids, and on the upper part between the bases of the autozooids. The ventral side of the rachis bears no autozooids, but on the dorsal surface there is little space left exposed. The autozooids are 10 mm. in length, and the tentacles 6 mm.

There are a few warty spicules of the usual three-edged shape.

Locality: Station 197; 9° 34′ 57″ N., 75° 36′ 30″ E.; 406 fathoms.

COMPARATIVE TABLE OF

Name.	Stalk.	Axis.	Rachis.
U. durissima, Kölliker.	Long, cylindrical, flexible, with a spindle-shaped enlargement at the lower end.	Nearly cylindrical.	A flat rhomboid expansion of the stalk.
U. güntheri, Kölliker.	Quadrangular, with many spindle-shaped enlargements.	Quadrangular.	Without inferior enlargement—bilateral.
U. thomsoni, Kölliker.	Quadrangular, with well-developed lower, but no upper enlargement.	Quadrangular, with excavated surfaces and rounded edges.	Indistinctly bilateral.
U. leptocaulis, Kölliker.	Very thin.	Quadrangular, with concave surfaces and blunt edges.	Bilateral.
U. simplex, Kölliker.	Cylindrical, thin, with a moderately thick sarcosoma.	Quadrangular, with concave surfaces and rounded edges.	Decidedly bilateral.
U. huxleyi, Kölliker.	With long enlargement below.	Indistinctly quadrangular.	Indistinctly bilateral.
U. carpenteri, Kölliker.	Enlargement at upper end, continuous, with rachis and long enlargement at lower end.	Quadrangular, with deeply excavated surfaces and rounded edges.	Indistinctly bilateral.

SPECIES OF UMBELLULA.

Autozooids.	Siphonozooids.	Spicules.	Notes.
Vary in number from three to nine, and have no definite arrangement; eight bands of spicules occur on the wall, and these are continued up the aboral surface of the tentacles.	Exceedingly numerous, minute; cover the whole of the prorachidial surface except a narrow median ridge, and extend in bands between the bases of the autozooids.	(*a*) Stalk—small rods or ellipses, with longitudinally arranged ribs formed of minute spine-like processes; (*b*) autozooids—(1) longer rods or ellipses, with fewer ribs; (2) long spindles, with rough jagged ends, and minute longitudinal ribs.	The definite arrangement of spicules in eight rows on the autozooid-wall does not seem to be of great specific moment. Depth—Laccadives, 1132 fathoms; "Scotia" specimens, 1742 fathoms; Japan, 565 fathoms.
Very large—alternate, 44 mm. long, with tentacles extended.	Very numerous, but leave a free line on both the pro- and the meta-rachidial surface.	Numerous in every part of cutis; needles three-edged, and granular at the end; 0·2–0·72 mm. in length, and 0·041–0·045 mm. in breadth.	Atlantic Ocean, a little N. of the Equator. Depth, 1850 fathoms.
Form a pendant bunch.	On pro- and meta-rachidial surfaces of the rachis; none on the stalk.	In all parts of sarcosoma.	Locality: North Atlantic Ocean, between Portugal and Madeira. Depth, 2125 fathoms.
Disposed alternately, 27–30 mm. in length.	Scanty—on rachis only.	In every part of the cutis. Those on the tentacles are disposed *longitudinally*; (*a*) stalk 0·11–0·18 mm. in length and 0·027–0·054 mm. in breadth; (*b*) tentacles 0·16–0·21 mm. long; (*c*) autozooid, 0·43–0·54 mm. long.	South-west of New Guinea. Depth, 2440 fathoms; near *U. güntheri*.
Alternate, small, rather hard, four in number.	None.	Numerous in every part of the cutis. Three-edged and granulated at their ends; 0·1–0·18 mm. long and 0·027–0·054 mm. broad.	Between San Francisco and Yeddo. Depth, 2050 fathoms. (Evidently young.)
Form a cluster at end of stalk, with trace of a bilateral arrangement, small, brown.	Numerous on whole stalk and between bases of autozooids, but none on the metarachidial side of the rachis between the polyps; all provided with a single tentacle.	None except in the end bulb of the stalk.	Locality: south of Yeddo. Depth, 565 fathoms.
Form a rosette at the end of the stalk, long, colourless.	Numerous on pro- and meta-rachidial sides of the rachis and along the whole stalk; all provided with one singly branched tentacle.	Only in the lowest part of the stalk.	South Polar Sea, south-west of Australia. Depth, 1975 fathoms.

COMPARATIVE TABLE OF

Name.	Stalk.	Axis.	Rachis.
U. magniflora, Kölliker.	Long swelling below, and a flattened and curved enlargement at its upper end.	Quadrangular, with concave surfaces and rounded edges.	No distinct rachis.
U. lindahlii, Kölliker.	Short, somewhat spirally-twisted, and bent in a large flat curve below the polyp-cluster.	Quadrangular, with rather deeply concave surfaces and ridge-like projecting edges.	In form of a polyp-cluster.
U. encrinus, Linn.	Upper part of stalk is twisted and bent in various ways; the whole stalk above the terminal swelling is turned like a spiral round its longitudinal axis.	Quadrangular, with deeply concave surfaces and rounded edges.	An apparently radiate hanging cluster.
U. geniculata, Studer.	Long, thin, slightly bent at the inferior end.	Quadrangular, surfaces concave.	Arises at a right angle. Autozooids arranged in a rosette—slightly bilateral.
Umbellula dura, n. sp.	Long, flexible, with an enlargement at lower end.	Almost cylindrical.	Short, and forms an irregular inverted cone flattened laterally.
U. intermedia, n. sp.	Long, slender, flexible, quadrangular in section; small enlargement at the base and another spindle-shaped higher up.	Quadrangular in section, with a groove on each side.	Bilateral, ovoid, with a ridge on the prorachidial surface.
U. rosea, n. sp.	Long, flexible, quadrangular in section, marked by a groove on each surface; two enlargements.	Quadrilateral and deeply grooved on the sides.	Short, bilateral.

SPECIES OF UMBELLULA—*continued.*

Autozooids.	Siphonozooids.	Spicules.	Notes.
Form a bunch at end of stalk without any trace of bilateral arrangement.	Numerous on the upper enlargement of the stalk at the base of the autozooids, and also on the long swelling of the stalk, and in its neighbourhood.	Absent.	South Sea, east of Kerguelen Island. Depth, 1600 fathoms.
Tentacles longer than the body. Autozooids 6–11 mm. long, tentacles about 20 mm.	Form tongue-shaped areas between the polyps. Below the cluster they are disposed all round, but downwards they pass into a single row on each side.	Absent except in the lower part of the stalk, where they are numerous; 0·016 mm. × 0·0064 mm.	This species includes *U. miniacea*, Lindahl; *U. pallida*, Lindahl; *U. gracilis*, Marshall; and *U. bairdii*, Verrill. Localities: Baffin's Bay, Davis Straits, Denmark Straits, Faroe Channel, North Greenland. Depths: 410, 1435, 568, 580-689, and 122 fathoms respectively.
Tentacles about same length as, or shorter than, the body.	Between the autozooids in tongue-shaped areas. Each has a ventral tentacle. Also over the whole stalk.	Very small spicules are found, but only in the lower end of the peduncle.	Danielssen links *U. magniflora*, Köll., with this species, but Jungersen keeps them as distinct species. Locality: Greenland and between Norway and the Faroes. Depth, 412–498 fathoms.
Tentacles are about 15 mm. long.	Occupy the space between the bases of the autozooids.	Absent.	China Sea. Lat. 21° 19′ N., long. 106° 24′ E. Depth, 680 fathoms.
Few and large, arranged in a whorl or bilaterally.	Small, not prominent; occur all over the rachis.	(*a*) Autozooids—(1) blunt rods covered by longitudinal ribs; (2) long rods with minute tubercles; (*b*) cœnenchyma—rods and quadriradiate forms.	Depth, 705 fathoms.
Five in number, one terminal; of medium length, with long tentacles; rigid, covered by spicules visible to the naked eye.	Small; occur in meshes in the spicular covering all over the rachis except on the prorachidial ridge.	(*a*) Rachis—rough rods or bars, blunt at the ends and ridged; (*b*) stem—smaller and proportionally wider rods.	This species is intermediate between *U. güntheri*, Köll., and *U. leptocaulis*, Köll. Depth, 1912 fathoms.
One terminal and two pairs of lateral; 5-6 mm. long, 2·5 mm. in diam., tentacles 6 mm. long.	On metarachidial extending to the prorachidial surfaces.	(*a*) Stalk — short, thick ellipses, with numerous ribs; (*b*) autozooids—a few longer and proportionally thinner, rough ellipses.	Depth, 1748 fathoms.

COMPARATIVE TABLE OF

Name.	Stalk.	Axis.	Rachis.
U. purpurea, n. sp.	Long, flexible, quadrangular, with prominent edges; two distinct enlargements.	Quadrangular in section, with sides deeply grooved.	Short and swollen; bilateral.
U. elongata, n. sp.	Large swelling at the base.	Quadrangular in section in lower part, circular or oval higher up.	Long, radially symmetrical.
U. köllikeri, n. sp.	Quadrangular to oval in section; slight swelling at junction with rachis.		Bilateral; prominent spine at origin of last autozooid.
U. radiata, n. sp.	Long; large swelling at lower end with four prominent ribs.	Roughly quadrangular in section.	Very long.
U. pendula, n. sp.	Very long, with large swelling at basal end.	Oval in section, yellow in colour.	Pendulous; bilateral in lower part, radial above.
U. indica, n. sp.	Tapers gradually from a large hollow terminal swelling with four "wings."	Quadrangular in section in stalk, circular in rachis.	Apparently radially symmetrical.
U. sp.	Cylindrical, thin, with a terminal swelling with four wing-like projections; peculiar bend or knee about half-way up.		Thin and wedge-shaped with a ridge.

SPECIES OF UMBELLULA—*continued.*

Autozooids.	Siphonozooids.	Spicules.	Notes.
Numerous; 42 mm. in length, with a diameter of 3 mm.	Numerous on the basal part of the rachis; prolonged into Λ-shaped points on the para- and meta-rachidial surfaces; completely cover the pro-rachidial surface.	Absent.	...
In short spirals; 18 mm. in length.	Occur all over the rachis and throughout the whole length of the stalk.	Restricted to lower part of stem; small rods smooth and rounded at ends; also a few four-cornered forms.	Depth, 360 fathoms.
Five in number arranged bilaterally.	At base of autozooids; middle line free.	Autozooids : — triangular in section, granular.	Depth, 1803 fathoms.
Numerous, arranged in whorls, long and slender.	Occupy the whole surface of the rachis.	Absent.	Depth, 490 fathoms.
Numerous, in irregular whorls; 40 mm. in length.	Abundant; occur all over the rachis except a short ridge on the lower part of the prorachidial surface.	Absent.	Depth, 188-220 fathoms.
Long and slender; 48 mm. in length.	Cover the whole surface of the rachis; wart-like, oval in shape, 3·2 mm. high.	Absent.	Depth, 238-290 fathoms.
Apparently irregularly disposed; very long, with short tentacles.	Cover the lower part of rachis, and occur also at base of autozooids.	Few warty spicules of the three-edged type.	Depth, 406 fathoms.

Family *Anthoptilidæ.*

Anthoptilum murrayi, Kölliker.
„ *decipiens*, n. sp.

Anthoptilum murrayi, Kölliker.

To this species we refer two incomplete specimens without autozooids; the stronger of the two is 620 mm. in length, and on an average 4 mm. in breadth.

The axis is sub-cylindrical, radially lamellar, with two lateral canals. In the lower part it is more or less quadrangular.

The stalk is short, and has a long spindle-shaped swelling at the lower end.

The rachis is long, more or less cylindrical in shape, marked by a broad groove on the prorachidial surface, and by a narrow groove on the opposite side.

There are no autozooids left on the rachis, but there are hollow oblong pit-like markings arranged in oblique transverse rows of two or three. It is possible that the adjacent edges of the autozooids may have been fused, and so formed rudimentary pinnules. The siphonozooids cover the whole surface of the rachis not occupied by bases of the autozooids.

The cœnenchyma on the stalk is of medium thickness, but on the rachis it is very thick.

There is no trace of spicules in the rachis, but small oval, quadrangular, or incipiently quadriradiate forms (about 0·01 mm. in length) are present in the lower end of the stalk.

The specimen has a dull white colour, while that described by Kölliker was pale red with brown polyps and colourless stalk.

With reference to the absence of autozooids, it is interesting to notice that Professor J. D. F. Gilchrist supplied the following note to Professor Hickson with reference to *Anthoptilum grandiflorum*: "Some difficulty was experienced in preventing the polyps from being washed off by the motion of the preservative fluid owing to the roll of the vessel" (*Alcyonaria of the Cape of Good Hope*, Part II., 1904, p. 233).

If we are right in referring our specimen to *A. murrayi*, the occurrence of this species in the Indian Ocean is very interesting. It was recorded by the "Challenger" from Station 50, North Atlantic, south of Halifax, 48° 8′ N., 63° 39′ W.; also by Verrill from the east coast of North America (*American Journal of Science* (3), vol. xxviii. p. 220). On the European side it has been recorded from the Bay of Gascony, 45° 57′ N., 6° 21′ W.; and from the following four stations south of Iceland: Station 83; 62° 25′ N., 28° 30′ W. Station 40; 62° N., 21° 36′ W. Station 65; 61° 33′ N., 19° W.; and Station 47; 61° 32′ N., 13° 40′ W.

Locality: Station 104; 11° 12′ 47″ N., 74° 25′ 30″ E.; 1000 fathoms.

Anthoptilum decipiens, n. sp. Plate VII. fig. 8.

This species is represented by a long substantial colony arising from a strong basal expansion. The total length is 720 mm., and the stem tapers from 4–2 mm. in breadth.

The stalk is short and conical; it expands gradually from its junction with the rachis downwards, and has a large knob in the centre of the base. The base thus resembles a "tam o' shanter"; or, to put it in another way, the stalk ends in a large knob, but before reaching the knob it expands into a large collar-like fold.

The rachis is very long, and is covered with a large number of autozooids in oblique rows, from three to eight in each row. The autozooids are fused for a very short distance upwards from their base, thus producing distinct though rudimentary pinnules. These are placed at a very small angle to the long axis of the rachis, and the upper end reaches often to the pararachidial surface. Their arrangement is somewhat perplexing, as there is distinct overlapping. Two occur almost opposite one another, and a third distinctly overlaps their insertion.

The autozooids are very long (10–25 mm.), thin walled, and packed with reproductive bodies which shine through the thin walls and give the polyps an appearance suggestive of fish-roe. (See Plate VII. fig. 8.)

The siphonozooids occur on the pararachidial and metarachidial surfaces, extending on to the bases of the rudimentary pinnules in small somewhat triangular areas insinuated between two adjacent autozooids. They also occur on the prorachidial surface, with the exception of a narrow median strip.

The spicules of the cœnenchyma at the lower end of the stalk are very minute rods (0·0056 mm. in length), which may be collected together to form star-shaped groups.

There are no spicules on the rachis or on the autozooids.

The most noteworthy features of this species are:

1. The perfect though rudimentary pinnules.
2. The peculiar overlapping arrangement of the pinnules.
3. The shape of the basal expansion with its knob-like termination into which the end of the axis extends.

Locality: Station 284; 7° 55′ N., 81° 47′ E.; 506 fathoms.

Family *Funiculinidæ*.

Subfamily Funiculininæ:

Funiculina quadrangularis (Pall.) = *Leptoptilum gracile*, Kölliker.

Funiculina gracilis, n. sp.

Subfamily Stachyptilidæ.

Stachyptilum maculatum, n. sp.

Subfamily Funiculininæ.

Funiculina quadrangularis (Pallas) = *Leptotilum gracile*, Kölliker.

A broken colony, 358 mm. in length.

The axis is quadrangular in section, with deep grooves on three of the sides, in the lower part. It is 0·95 mm. in diameter, and tapers gradually to a delicate thread at the tip.

All the autozooids have fallen off, so that their arrangement cannot be ascertained with certainty, but they seem to have occurred alternately in a single row on each side of the axis. Several detached autozooids are 7 mm. in length and 2 mm. in breadth. Each has eight projecting apical points formed of a number of small smooth blunt rod-like spicules, 1·2 mm. in length, 0·038 mm. in diameter.

The autozooids are colourless, but the dark brown contents of the stomodæum are seen through the walls.

We see no sufficient reason to separate this form from *Leptoptilum gracile*, though the autozooids in Kölliker's specimens were only 3–4 mm. in length. Grieg and Jungersen have shown that *Leptoptilum gracile* is simply a stage in the life history of *Funiculina quadrangularis*.

Locality: Bay of Bengal; 753 fathoms. Previously recorded as *Leptoptilum gracile* from New Zealand. *Funiculina quadrangularis* is recorded from North Sea, the Atlantic Ocean (on the European and American sides), the Mediterranean, etc.

Another delicate graceful specimen is referred to this species. It is complete. and has a total length of 49 mm., of which the stalk occupies 22 mm.

The colony is extremely delicate, the rachis differing from the stalk only by the presence of the autozooids.

The autozooids are arranged in twos almost at the same level, or one slightly above the other. They lie partly on the pararachidial, partly on the metarachidial surface. They stand almost at right angles to the surface of the stalk, or are directed slightly upwards. Between each pair of autozooids two small zooid-like structures are present, but these are undoubtedly young stages of autozooids.

The polyp-calyces have eight projecting points, each composed of a number of spicules. On the bodies of the calyces the spicules are arranged in an irregular manner, but mostly transversely; underlying the transverse spicules there are eight narrow bands of longitudinally arranged spicules. These bands are formed in the following manner: each point divides into two sets of diverging spicules, each set then joins with the adjacent set of the next point, and the two sets thus united form a band extending towards the base of the calyx.

The calyces are from 2–3·5 mm. in height, and some of the younger autozooids between the others are from 1–1·5 mm. in length. Their walls are very thin and translucent.

The cœnenchyma is thin and transparent, allowing the axis to shine through.

The spicules of the polyp-calyces and cœnenchyma are smooth fluted rods, perfectly transparent. Those of the stem are on an average 0·22 mm. in length by 0·015 mm. in breadth; those of the calyx wall are from 0·4 to 0·5 mm. in length by 0·15 mm. in breadth.

The colony is colourless, except for the stomodæum, which is blackish-brown in colour.

This specimen agrees closely with the specimen described by Kölliker.

Locality: Station 235; 14° 13′ N., 93° 40′ E.; 370–419 fathoms.

Funiculina gracilis, n. sp. Plate VII. fig. 10; Plate IX. fig. 4.

A large number of specimens, mostly broken and slightly damaged, are placed in this species. A representative example was 350 mm. in length, with a breadth of rachis about 1·25 mm.

It consists of a stalk (100 mm.) and an elongated quadrilateral rachis. There is no difference between the stalk and rachis, apart from the presence of the autozooids.

The axis is quadrangular with hollow sides, thus giving the stem, which varies from 0·7–1·5 mm. in diameter, a quadrangular grooved appearance. Towards the basal end the stem diminishes and then swells, and forms a somewhat club-shaped enlargement which is usually slightly curved.

The rachis is quadrilateral and long, having the autozooids arranged in groups, each group consisting of two irregular rows of three to four polyps.

The calyces are produced into eight sharp teeth—the tips of bands of spicules. On the upper part of the calyx, the spicules lie for the most part longitudinally, but towards the basal portion they are more irregular and are often obliquely transverse. Numerous siphonozooid-like bodies, white in colour in marked contrast to the dull grey of the autozooids, occur among the bases of the latter, principally in a somewhat broken single line on the metarachidial surface. They are furnished with eight short lobe-like projections at the tip, and rise to a height of 1 mm.

The height of the autozooid calyces varies from 3–5 mm., and the diameter from 1–1·5 mm.

The colour of the rachis is yellowish to whitish, the basal portion of the calyces is also whitish, while the upper portion is bluish-black to a dull grey.

The ova are fairly abundant, and *embryos* were found with a slightly brown outer coating and a yellow interior with a large cavity.

The spicules are long slender fluted rods, blunt or pointed at the ends, and often slightly curved at one end. The following measurements were taken of length and breadth in millimetres:

0·8×0·04; 0·7×0·035; 0·4×0·02.

It is possible that the form which we have identified with *Leptoptilum gracile*, Kölliker, and referred to *Funiculina quadrangularis*, Pallas, is really the young stage of this new species of *Funiculina*.

Locality: Station 197; 9° 34′ 57″ N., 75° 36′ 30″ E.; 406 fathoms.

Subfamily Stachyptilidæ.

Stachyptilum maculatum, n. sp. Plate VII. figs. 5 and 9; Plate IX. fig. 16.

This species is represented by one complete club-shaped specimen, 62 mm. in height.

The rachis is club-shaped, gradually increasing in diameter and reaching its maximum a little below the apex. It presents a rather striking appearance owing to the presence of brown siphonozooids on the surface not occupied by autozooids.

The autozooids are arranged in ten oblique rows on each side, each row consisting of 3–4 polyps. They are fairly large, and capable of complete retraction; they show no spicules in their thin walls. The tentacles are of median length, and gradually taper to the distal end. They have one row of 12–14 short blunt pinnules on each side.

The siphonozooids form slight brown projections on the surface, and appear as eight-rayed stars. They vary in diameter from 0·2–0·3 mm., and the larger individuals are slightly oval. They cover all the surface of the rachis not occupied by the autozooids, and seem also to occur in two grooves that run up the rachis. These grooves on the rachis may be due to contraction in the preserving fluid.

The spicules are rod-like, with jagged ends, or four- to five-rayed forms. They vary in length from 0·1–0·16 mm., and in breadth from 0·025–0·04 mm.

This species differs from those previously described in the greater variety of spicules, in having more of the rachis covered, and in the absence of papillæ at the upper end of the stalk.

Locality: Station 213; 21° 25′ N., 68° 02′ 30″ E.; 137–131 fathoms.

Family *Virgularidæ* (*Pavonaridæ*, Jungersen).

Pavonaria willemoësii, Kölliker = *Microptilum willemoësii*, Kölliker.

A broken specimen with a total length of 116·5 mm., of which the stalk occupies 33 mm., thus leaving a long rachis. The axis is cylindrical, from 0·35–0·4 mm. in diameter. The diameter of the stalk is 0·4–0·5 mm., and it is cylindrical just below the origin of the rachis. The cœnenchyma is thin, and allows the axis to shine through on the prorachidial surface, but is considerably thicker on the metarachidial surface.

The thirty-six autozooids are arranged alternately in one row on each side of the rachis. They vary from 2–3·2 mm. in height. Between these there are some smaller polyps interspersed, and there are 3–4 small polyps below the lowest well-developed autozooid. The calyces are prominent, furnished with two strong points on the abaxial side which are separated from the axial side by a deep indentation. The spicules are arranged longitudinally on the abaxial and axial surfaces, but on the latter they tend to be placed transversely at the edges. On the lateral surfaces the spicules are fewer in number, and some are directed at right angles to the stalk, so that they cross the others which are placed longitudinally.

The spicules are in the form of needles, varying from 0·5–1 mm. in length on the rachis and the autozooids, but on the stalk they are very small (0·12–0·15 mm.).

On the small immature polyps the spicules are arranged as on the mature forms, thus producing a small conical projection.

No siphonozooids are distinguishable. Jungersen seems to have shown conclusively that *Microptilum* is simply a young stage of *Pavonaria*; and as our specimen agrees closely with *Microptilum willemoësii*, Kölliker, we have recorded it under the title *Pavonaria willemoësii.*

Locality: Andaman Sea; 650 fathoms.

Geographical Distribution.—This genus is widely distributed; *P. finmarchica* has been recorded from the coast of Norway, from the vicinity of Iceland, and from the east coast of North America, at depths varying from 60–980 fathoms; *P. africana* (Studer) from the Atlantic and the West Coast of Africa, 10° 12·9′ N., 17° 23·5′ W.; *P.* sp. from the Japanese Seas and the Gulf of Korea, and *Microptilum willemoësii* from Station 235, south of Yeddo, 34° 7′ N., 138° E., at a depth of 565 fathoms. Another species has been recorded from Behring Island, and two more, *P. dofleinii* and *P. californica*, from the Californian coast.

Family *PENNATULIDÆ.*

Subfamily Pennatulinæ.

Pennatula indica, n. sp.

" *veneris*, n. sp.

" *splendens*, n. sp.

" *pendula*, n. sp.

Subfamily Pteroëididæ.

Pteroëides triradiata, n. sp.

Subfamily Pennatulinæ.

Pennatula indica, n. sp. Plate VIII. fig. 1.

Several beautiful specimens, the largest of which (A and B) are 150 and 105 mm. in length.

The stalk is long and straight, whitish in colour, with a pinkish tinge. Its length is about half that of the entire colony (64 mm. in A and 49·5 mm. in B). It shows a number of swellings, three in B, the smaller, and four in A, the larger specimen. There is a very small swelling near the base, a second slightly bigger, and almost touching the first, a third about the middle point of the stalk, and in the largest specimen (A), a fourth swelling at a point three-fourths the length from the base.

The following measurements were taken :

Specimen.	A.	B.	C.	D.
Total length	150 mm.	105 mm.	84 mm.	78 mm.
Length of peduncle	64 mm.	49·5 mm.	34 mm.	37 mm.
Number of pinnules	29/30	17/19	21/21	15/15
Length of well-developed pinnule	35 mm.	23·2 mm.	17 mm.	22 mm.
Breadth of same at base	4·2 mm.	3·2 mm.	2·5 mm.	2 mm.
Number of polyps on same	11	11	7	7

Owing to the contraction of the cœnenchyma the tip of the axis projects at the basal end. It is thin and thread-like, and is coiled several times into a sort of spiral.

The crimson-lake colour of the rachis and pinnules presents a striking contrast to the whitish or pinkish colour of the stalk. The lower pinnules, varying in number from 3–5, are small and rudimentary in all the specimens, the lowest being recognisable only as a band of slightly darker spicules lying across the sides of the rachis. The fully developed pinnules are long, narrow, sword-shaped, with calyces on their edges. They are covered by long spicules arranged parallel to the length and closely packed together, forming a complete casing. The calyces stand not at right angles to the pinnule, but at a small angle, and have their apices directed towards the distal end. Each calyx is formed of longitudinally disposed spicules similar to those of the pinnule, and has its apex divided into eight points, each of which is made up of at least two spicules. The calyces reach a height of 4·1 mm. The autozooids are capable of complete retraction within the calyces. Each tentacle has on its aboral surface a broad band of spicules which are arranged more or less longitudinally.

Along the prorachidial surface there is a broad band free from autozooids. This band is divided into two smaller bands by a long narrow furrow which extends the whole length of the rachis. The band is closely covered by numerous siphonozooids, each of which has a small calyx which is closely appresssd to the surface of the rachis, and thus appears only as a platform of spines lying at an acute angle to the surface, with the apex directed towards the tip of the rachis. On the metarachidial surface siphonozooids are also present, forming a single row on each side of the middle line. They are not so closely pressed against the surface. The general cœnenchyma is fairly thick, and has abundant spicules. The spicules are long slender fluted rods, varying from 0·15–3 mm. in length, and from 0·01–0·13 mm. in width.

Both ova and *embryos* were found in the pinnules.

Localities: Station 260; 8° 28′ 15″ N., 76° 07′ E.; 487 fathoms. Station 323; 16° 25′ N., 93° 43′ 30″ E.; 463 fathoms. Station 230; 7° 40′ N., 70° 00′ 52″ E.; 824 fathoms.

Pennatula veneris, n. sp. Plate VIII. fig. 8.

Three magnificent specimens, with a beautiful colour scheme. The stalk and rachis are dark red, the pinnules are transparent, the autozooids are red with white tentacles.

The short stalk ends in a swollen base, and just below its junction with the rachis there is a short thick spindle-shaped enlargement.

The rachis, which is many times longer than the stalk, tapers gradually towards the tip, where it becomes thread-like and loses itself in the base of a small pinnule. On the metarachidial surface a deep broad groove extends the whole length; the prorachidial surface tends to be rounded.

The pinnules are numerous, and are placed on the pararachidial surfaces, and present a striking contrast to the darker coloured rachis. They are placed almost parallel to the axis of the rachis, and show considerable variation in size, the best developed pinnules being about half-way up the rachis. A well-developed pinnule is triangular in shape with a broad base, and one side, the prorachidial edge, is slightly curved, while the metarachidial edge bears the autozooids. In the younger stages of the pinnules the terminal autozooid is very long, and separated from the rest by a considerable distance, and the peculiar appearance thus resulting is accentuated in still younger stages by the whole pinnule being whip-like, and even resembling a flattened thread with only one, the terminal, autozooid. The pinnules are transparent, and contain abundant ova.

The autozooids stand in a single row on the edge of the pinnule. They have short stalks, and are expanded at the oral end into a somewhat bell-like shape. There are no projecting spines. The colouring of the polyp is striking. The short stalk is translucent in the lower part, then a brownish-yellow tinge begins to appear, the brown colour becomes gradually intensified in the body of the polyp until just below the origin of the tentacles, where it becomes a deep red. The body wall is marked by eight lines. Towards the base of the pinnule the polyps decrease in size, and finally appear as a row of small white warts on the edge. The tentacles are short, and have one row of pinnules on each side. On the aboral surface they have a narrow band of small red spicules extending up the middle line to the tip. The tentacles curve inwards when at rest, and then they appear like a small white dome marked by eight bands of red which radiate from the apex.

The siphonozooids occur principally on the pararachidial surfaces in long narrow bands between the pinnules, the upper end of each band reaching for a short distance on to the metarachidial surface, and the lower end for a similar distance on to the opposite side. The middle of each band of zooids is marked at the middle point by a slight broadening in the upper part of the rachis, or by a slight S-shaped twist in the lower part of the rachis.

The following measurements were taken:

Specimen.	No. I.	No. II.	No. III.
Length of specimen	243 mm.	221 mm.	186 mm.
Length of peduncle	46 mm.	40 mm.	37 mm.
Number of pinnules	32/33	27/27	24/23
Length of well-developed pinnule	26·5 mm.	19 mm.	16 mm.
Breadth of same near base	10 mm.	10 mm.	10 mm.
Number of autozooids in same	29+7	26+5	22+8

(The unit after the plus sign in the number of autozooids indicates the number of immature forms towards the base of the pinnules.)

In specimen No. II. the terminal autozooid of a pinnule may reach a length of 9 mm.

The spicules of the autozooids are all of one shape, flattish bars rounded at the ends and constricted at the middle. They are from 0·04 to 0·045 mm. in length by 0·02 to 0·015 in breadth. They vary in colour from brownish to red.

The spicules of the stem are similar in form to those of the autozooids, and of similar dimensions (0·04×0·02).

Locality: N.-W. of Calicut; 100 fathoms.

Pennatula splendens, n. sp. Plate VIII. fig. 5.

The short stalk ends in a slight swelling, and is somewhat curved. Just below the rachis there is a spindle-shaped enlargement, which has a diameter of 7 mm.

The rachis (272 mm.) is four times longer than the stalk (68 mm.), dark red in the lower part, and whitish to greyish in the upper part. On the metarachidial surface there is a groove; the prorachidial surface tends to be rounded.

The numerous pinnules lie almost parallel to the long axis of the rachis, on the pararachidial surfaces. They are triangular, with a greatly elongated base. The pinnules are somewhat varied in shape at the different parts of the rachis, much smaller and narrower in the lower part, but never showing any trace of the whip-like development seen in *Pennatula veneris*. In colour also they show variation, the younger pinnules being similar in colour to the stalk, while the better developed pinnules are greyish to dirty white.

The autozooids are densely crowded on the metarachidial edge of the pinnule. They are somewhat bell-shaped, and stand on a short peduncle, which in the case of the autozooids on the upper pinnules passes through the same colour scheme as in those of *Pennatula veneris*, but in the lower pinnules the peduncles are uniformly dark red. There are no projecting spines or points in the autozooids, which are arranged so closely that they appear to be in two or three rows. The tentacles are short, white in colour, and have no band of red spicules on the aboral surface.

The siphonozooids form broad bands which extend upwards between the pinnules on the pararachidial surfaces, but their lower ends extend so far down as to form a distinct band on the prorachidial margin of the lateral surfaces.

The following measurements were taken :

> Breadth of rachis near base, 3 mm. ; near tip, 1·8 mm.
> Length of well-developed pinnule, 23–25 mm.
> Breadth, 12 mm. ; about 6 mm. near base and tip.
> Number of autozooids, 51 + 10 (19 near base, 45 near tip).

The spicules of the pinnules and autozooids are somewhat flattened ovals or ellipses, brownish in colour ; the following measurements were taken of length and breadth in millimetres :

$$0{\cdot}04 \times 0{\cdot}02\ ;\ 0{\cdot}03 \times 0{\cdot}04\ ;\ 0{\cdot}029 \times 0{\cdot}04.$$

They are not so much constricted at the middle as in *Pennatula veneris.*

This species differs from *Pennatula veneris* in the following details, although they show a remarkable superficial resemblance.

1. The lower pinnules are different in shape, not having any trace of the whip-like development of the terminal autozooid.
2. The colour of the pinnules is different, the lower being dark red, the upper and more developed being white to greyish.
3. There are no bands of red spicules on the aboral surfaces of the tentacles.
4. The lower ends of the siphonozooid-bands fuse and form a continuous band or strip on the prorachidial margin of the pararachidial surfaces.

Locality : Station 239 ; 11° 49′ 30″ N., 92° 55′ E. ; 55 fathoms.

Pennatula pendula, n. sp. Plate VII. fig. 1; Plate VIII. figs. 7 and 10.

This species, which in some ways resembles *P. murrayi*, Kölliker, is represented by numerous specimens.

The usual colour of the stalk and rachis is dark red, but in some the rachis is pinkish below, then almost pure white, and then coral red. The pinnules are usually white, but some are coral red. The autozooids are red, their tentacles white, the siphonozooids are yellowish.

The axis is cylindrical above, quadrangular below; at the lower end two of the angles are produced into wing-like expansions.

The translucent pinnules, of which there may be over fifty pairs, are elongated triangles or lanceolate. The calyces have eight projecting spines.

The siphonozooids are in two rows on the pararachidial surfaces.

As there are many specimens, we give several sets of measurements to show quantitative differences.

Length of Whole.	Length of Stalk.	Breadth of Enlargement of Stalk.	Length of Rachis.	Breadth of Rachis.	Breadth of Base of Pinnules.	Polyps on Pinnules.
256 mm.	100 mm.	3 mm. Breadth of stalk, 1·2 mm.	156 mm.	Near base, 1·5 mm. Near tip, 1·8 mm.	Near base, 2·9 mm. Near tip, 7 mm. About half-way up, 5·5 mm.	Near base, 8. Near tip, 11. About half-way up, 10.
229 mm.	50 mm.	3 mm. Breadth of stalk, 1·4 mm.	179 mm.	Near base, 2·8 mm. Near tip, 1·5 mm.	Near base, 2·6 mm. Near tip, 3 mm. About half-way up, 4 mm.	Near base, 7. Near tip, 10–11. About half-way up, 9.
240 mm.	122 mm.	2 mm. Breadth of stalk, 1·2 mm.	118 mm. 53 pairs of pinnules.	Near base, 2 mm. Near tip, 1·5 mm.	Near base, 4 mm. Near tip, 3·5 mm. About half-way up, 5·2 mm.	Near base, 9. Near tip, 7. About half-way up, 10.

Localities: Andamans; 238–290 fathoms. Station 235; 14° 13′ N., 93° 40′ E.; 370–419 fathoms. Station 240; 11° 32′ N., 92° 46′ E.; 194 fathoms.

Subfamily Pteroëididæ.

Pteroëides triradiata, n. sp.

At the base of the stalk there is a slight swelling, 8 mm. in length and 3 mm. in breadth, from the end of which the bare axis projects and ends in a hook-shaped portion.

At a distance of 15 mm. from the end of the stalk a large swelling (4·8 mm.

in breadth) arises and extends up the stalk, gradually tapering till the stalk has reached its normal thickness at the origin of the lowest pinnules.

The rachis is nearly 30 mm. in breadth. The bare prorachidial surface is 2·3 mm. across. The *three* large and well-developed rays in the pinnules give the specimen a very stiff and spiny appearance. The pinnules may be said to be sword-shaped, and vary in width from 2–3 mm. The prorachidial edge is occupied by the strong first ray, which consists of about three long spicules. The presence of only three rays is a distinctive feature of this species.

The following measurements were taken of the two specimens:

Specimen.	Stalk.		Rachis.		Pinnules.		Number of Rays.	Zooid Streak.	Zooid Plate.
	Length.	Breadth.	Length.	Breadth.	Length.	Width.			
A	30 mm.	2 mm.	33 mm.	24 mm.	15 mm. (16 pairs of pinnules.)	2·8 mm.	3 at most.	Not seen.	Marginal (?).
B	39 mm.	3·7 mm.	67·6 mm.	24 mm.	17 mm. (23 pairs of pinnules.)	3 mm.	3 at most.	Not seen.	Marginal (?).

Locality: Station 259; 10° 08′ 43″ N., 75° 33′ 30″ E.; 56 fathoms.

REFERENCES TO LITERATURE.

1888. Agassiz, A.—Characteristic Deep-Sea Types—Polyps. Three Cruises of the "Blake," vol. ii. pp. 142–148.

1862. Alder, J.—Descriptions of some new and rare Zoophytes of the Coast of Northumberland. Ann. Mag. Nat. Hist. 3 ser. ix.

1898. Ashworth, J. H.—The Stomodæum, Mesenterial filaments and Endoderm of Xenia. Proc. Roy. Soc. London, lxiii. pp. 443–446.

1899. Ibid.—The Structure of Xenia hicksoni, n. sp., with some Observations on Heteroxenia elisebethæ Kölliker. Quart. Journ. Micr. Sci. xlii. pp. 245–304.

1900. Ibid.—Report on the Xeniidæ collected by Dr. Willey. Willey's Zool. Results, pt. iv. pp. 509–530, pls. lii. and liii.

1834. Blainville.—Manuel d'Actinologie ou de Zoophytologie.

1862. Bleeker, P.—Sur les espèces nouvelles de Pennatulina, etc. Rev. et Mag. Zool. 2 ser. xiv. pp. 38–42.

1862. Ibid.—Naturkund. Tijdschr. Batav. xx.

1761. Bohadsch.—De quibusdam Animalibus Marinis, p. 112 (Funiculina).

1900. Bourne, G. C.—A Treatise of Zoology. Edited by E. Ray Lankester. Part ii. The Anthozoa, pp. 1–80.

1895. Ibid.—On the Structure and Affinities of Heliopora cœrulea (Pallas), with some observations on Xenia and Heteroxenia. Phil. Trans. clxxxvi.

1900. Ibid.—On the genus Lemnalia, also branching systems of Alcyonacea. Trans. Linn. Soc. London, 2 ser. vii. pp. 521–538.

1871. Brüggemann, F.—Corals and Polyps of the West Coast of America. Amer. Naturalist, v. pp. 306 and 307.

1879. Ibid.—An account of the Petrological, Botanical, and Zoological Collections made in Kerguelen's Land and Rodriguez: Corals. Phil. Trans. clxviii.

1896. Brundin, J. A. Z.—Alcyonarien aus der Sammlung der Zool. Mus. in Upsala. Svenska Vet.-Akad. Handl. xxii. pt. iv. No. 3, pp. 1–22, 2 pls.

1902. Bujor, P.—Sur l'organisation de la Vérétille (Veretillum cynomorium (Pall.) Cuv. var. stylifera Köll.). Arch. Zool. exp. Notes (3), ix. pp. 49–50, 7 figs.

1898. Burchardt, E.—Alcyonaceen von Thursday Is. und von Amboina. Denkschr. med.-nat. Ges. Jena, pp. 656–668, pls. liv.–lvii.

1880. Carter, H. J.—Specimens from Gulf of Manaar. Ann. Mag. Nat. Hist. 5 ser. vol. v.

1813. Cavolini, Ph.—Abhandlungen über Pflanzenthiere des Mittelmeeres. Uebersetzt von W. Sprengel. Nürnberg.

1903. Chun, C.—Aus dem Tiefen des Weltmeeres. 2te Aufl.

1817. Cuvier.—Règne Animal.

1846. Dana, J. D.—Zoophytes. United States Exploring Expedition. Philadelphia, vii.

1859. Danielssen, D.—Forh. Vidensk. Selsk. Christiania, p. 251.

1887. Ibid.—The Norwegian North-Atlantic Expedition, 1876–1878. Vol. v. Alcyonida, pp. 1–147, 23 pls.

1896. Dendy, A.—On Virgularia gracillima in Lyttleton Harbour. Trans. New Zealand Institute, xxiv. pp. 256–257.

1861. Duchassaing et Michelotti.—Memoire sur les Coralliaires des Antilles. Mem. R. Acad. Sci. Torino, 2 ser. xix. pp. 279–365, 10 pls.

1866. Ibid.—Supplement, *ibid.* xxiii. pp. 97–206, 11 pls.

1832. Ehrenberg, C. G.—Beiträge zur Kenntniss der Korallenthiere des rothen Meeres. Abh. Akad. Wiss. Berlin, 1832, p. 284.

1834. Ibid.—Die Korallenthiere des rothen Meeres. Berlin, p. 156.

1876. Eisen.—Bidrag til Kännedomen on Pennatulidslagtet Renilla. Svenska Vet.-Akad. Handl. xiii.

1764. Ellis, J.—Phil. Trans. liii. pp. 423–425.

1786. Ellis and Solander.—The Natural History of many curious and uncommon Zoophytes. London.

1797. Esper, E. J.—Die Pflanzenthiere in Abbildungen nebst Beschreibungen. Nürnberg, 1791–1797, 4 vols.

1875. Fisher, Wm. I.—On a new species of Alcyonoid Polyp. Proc. Calif. Acad. Nat. Sci. v. p. 418, 1 pl.

1844. Forbes.—Ann. Mag. Nat. Hist. xiv. pp. 413–414 (Funiculina).

1847. Ibid.—Johnston's History of British Zoophytes. 2nd edition.

1775. Forskål.—Descriptiones animalium quæ in itinere orientali observavit Havniæ.

1888. Fowler, G. H.—On a new Pennatula from the Bahamas. Proc. Zool. Soc. London, pp. 135–140, pl. vi.

1894. Ibid.—On two Sea-pens of the family Veretillidæ from the Madras Museum. Proc. Zool. Soc. London, pp. 376–379, pl. xxii.

1863. Gabb, Wm. M.—Description of two new species of Pennatulidæ from the Pacific Coast of the United States. Proc. Calif. Acad. Nat. Sci. ii. pp. 166–167.

1864. Ibid.—Description of a new species of Virgularia (gracilis) from the Coast of California. Proc. Calif. Acad. Nat. Sci. iii. pt. ii. p. 120.

1867. Genth, C.—Ueber Solenogorgia tubulosa (eine neue Gattung der Gorgoniden). Zeitschr. f. wiss. Zool. xvii. pp. 429–442, 3 pls.

1896. Germanos, N. K.—Gorgonaceen von Ternate. Abh. Senck. Nat. Ges. xxiii. Heft 1, pp. 145–184, 4 pls.

1835. Gray, J. E.—Proc. Zool. Soc. London, 1835.

1851. Ibid.—Acanthogorgia hirsuta. Proc. Zool. Soc. London, pl. iii. fig. 2.

1857. Ibid.—(1) Characters of a new genus of Corals (Nidalia); (2) Description of a new genus of Gorgonidæ (Acanthogorgia). Proc. Zool. Soc. London, xxv.

1859. Ibid.—Description of some new genera of Lithophytes. Proc. Zool. Soc. London, xxvii. pp. 479–486.

1859. Ibid.—On the arrangement of Zoophytes with pinnated tentacles. Ann. Mag. Nat. Hist. 3 ser. iv.

1860. Ibid.—Description of a new Coral from Madeira. Proc. Zool. Soc. London, xxviii. pp. 393–394, 1 pl.

1860. Ibid.—Ann. Mag. Nat. Hist. 3 ser. vii. pp. 214–215.

1862. Ibid.—(1) Description of some new species of Spongodes and of an allied genus Morchellana in British Museum. Proc. Zool. Soc. London, pp. 27–31; (2) Description of two new Genera of Zoophytes (Solenocaulon and Bellonella), pp. 34-37; (3) Notes on some specimens of Claviform Pennatulidæ (Veretilleæ) in British Museum, pp. 31–34.

1862. Ibid.—Notice of a second species of Paragorgia from Madeira. Ann. Mag. Nat. Hist. 3 ser. x.

1865. Ibid.—Notice on Rhodophyton, a new genus of Alcyonidæ. Proc. Zool. Soc. London, pp. 705–707.

1867. Ibid.—Additional note on Corallium johnsoni. Proc. Zool. Soc. London, pp. 125–127.

1868. Ibid.—(1) Note on a new Japanese Coral (Isis gregorii), pp. 263–264; (2) Descriptions of some new genera and species of Alcyonoid Corals in the British Museum, pp. 441–445. Ann. Mag. Nat. Hist. 4 ser. ii.

1869. Gray, J. E.—(1) Notes on Fleshy Alcyonoid Corals; (2) Siliceous spicules in Alcyonoid Corals, p. 96. *Loc. cit.* iii. pp. 21–25 and pp. 117–131.

1870. Ibid.—Notes on some new genera and species of Alcyonoid Corals in the British Museum. *Loc. cit.* v. pp. 405–408.

1870. Ibid.—Catalogue of the Lithophytes or Stony Corals in the Collection of the British Museum, London.

1870. Ibid.—Catalogue of Sea-Pens or Pennatulariidæ in the Collection of the British Museum, London.

1872. Ibid.—Alcyonoid Corals and Sponges from the Gulf of Suez. Ann. Mag. Nat. Hist. 4 ser. ix. pp. 481–482.

1872. Ibid.—The Clustered Sea Polyp Umbellula. *Loc. cit.* 4 ser. x. p. 151.

1872. Ibid.—Notes on Corals from South and Antarctic Seas. Proc. Zool. Soc. London, pp. 744–747, 3 pls.

1882. Ibid.—Corallium. *Loc. cit.* p. 223.

1869. Gray and Carter.—On Spongodes conglomeratus and a new genus of Fleshy Alcyonoids (Euslerides). Ann. Mag. Nat. Hist. 4 ser. iv. pp. 360–361.

1887. Grieg, J. A.—Bidrag til de Norske Alcyonarier. Bergen's Museum, Aarsber. p. 126, 9 pls.

1888. Ibid.—To nye Cornularier fra den Norske-Kyst. Bergen's Museum, Aarsber.

1891. Ibid.—Oversigt Norges Pennatalider. Bergen's Museum, Aarsber. No. i. p. 118.

1863. Grube, E.—Description of a new Coral (Lithoprimnoa arctica). Ann. Mag. Nat. Hist. 3 ser. xii. pp. 166–167.

1886. Haacke, W.—Zur Physiologie der Anthozoen. Zool. Garten, xxvii. p. 284 et seq.

1875. Haeckel, E.—Arabische Korallen, Berlin.

1900. Hargitt and Rogers.—The Alcyonaria of Portorico. U.S. Fish. Comm. Bull. vol. ii.

1900. Hedlund, T.—Einige Muriceiden der Gattungen Acanthogorgia, Paramuricea, and Echinomuricea in Zoologischen Museum der Univ. Upsala. Svenska Vet.-Akad. Handl. xvi. pt. iv. No. 6, pp. 1–18, 3 pls.

1883. Herdman, W. A.—On the Structure of Sarcodictyon. Proc. Roy. Phys. Soc. Edinburgh, vol. viii.

1683. Herklots, J. A.—Description de deux espèces de Pennatuliden des mers de la Chine. Nederl. Tijdschr. Dierk. 1 Jaarg. pp. 31–34.

1858. Ibid.—Polypiers nageurs ou Pennatulides, p. 30, 7 pls.

1883. Hickson, S. J.—On the Ciliated Groove (Siphonoglyphe) in the Stomodæum of Alcyonarians. Proc. Roy. Soc. London, xxxv. No. 226; also Phil. Trans. clxxiv. pt. iii.

1883. Ibid.—Quart. Journ. Micr. Sci. xxiii. p. 556 (Tubipora).

1894. Ibid.—A Revision of the Genera of the Alcyonaria Stolonifera with one new genus and several new species. Trans. Zool. Soc. London, xiii. p. 325.

1895. Ibid.—Anatomy of Alcyonium digitatum. Quart. Journ. Micr. Sci. xxxvii. p. 343.

1900. Ibid.—Alcyonaria and Hydrocorallinæ of the Cape of Good Hope. Part i. Marine Investigations in South Africa, i. No. 5, pp. 67–96, 6 pls.

1901. Ibid.—Alcyonium. Mem. Liverpool Marine Biol. Committee. London.

1902. Ibid.—Alcyonaria. Southern Cross Collection (Alcyonium pæsseleri), p. 293.

1902. Ibid.—Report of the Southport Meeting of the British Association (Polymorphism in Pennatula murrayi). Science, lxviii. p. 614.

1904. Ibid.—The Alcyonaria of the Cape of Good Hope. Part ii. Marine Investigations in South Africa, vol. iii. pp. 211–239, 3 pls.

1903. Ibid.—Alcyonaria of the Maldives. Fauna and Geogr. Maldives and Laccadives, ii. pt. 1, pp. 473–500, pls. xxvi. and xxvii.

1903. Ibid.—*Loc. cit.* ii. pt. 4, pp. 807–826, pl. lxvii.

1905. Hickson, S. J.—On a new species of Corallium from Timor. K. Akad. Wetensch. Amsterdam.

1900. Hickson, S. J., and Hiles, Isa L.—The Stolonifera and Alcyonacea collected by Dr. Willey in New Britain, etc. Willey's Zool. Results, pt. iv. pp. 493–508, 2 pls.

1875. Higgins, H. H.—Korallen in Sicilien. Natur. lxvi. (N. F. xxxiv.) p. 268.

1879. Ibid.—The Red Coral of Commerce—Corallium rubrum. Proc. Lit. Phil. Soc. Liverpool, No. 33, pp. xlviii-li.

1899. Hiles, Isa L.—The Gorgonacea collected by Dr. Willey. Willey's Zool. Results, pt. ii. pp. 195–206, 2 pls.

1899. Ibid.—Report on the Gorgonacean Corals collected by Mr. J. Stanley Gardiner at Funafuti. Proc. Zool. Soc. London, pp. 46–52, 4 pls.

1895. Holm, O.—Beiträge zur Kenntniss der Alcyonidengattung Spongodes. Zool. Jahrb. viii. pp. 8–57, 2 pls.

1901. Ibid.—Weiteres über Nephthya und Spongodes. Results of the Swedish Zoological Expedition to Egypt and the White Nile.

1860. Horn, G. H.—Descriptions of three new species of Gorgonidæ. Proc. Acad. Nat. Sci. Philadelphia, p. 233.

1885. Hubrecht, A.—On a new Pennatulid from the Japanese Sea (Echinoptilum macintoshii). Proc. Zool. Soc. London.

1838. Johnston, G.—A History of British Zoophytes, Edinburgh, 2nd ed. 1847.

1861. Johnson, J. Y.—Description of a second species of Acanthogorgia from Madeira. Proc. Zool. Soc. London, pp. 296–298.

1861. Ibid.—Notes on the Sea Anemones of Madeira (Cornularia atlantica, n. sp.). *Loc. cit.* p. 298.

1862. Ibid.—Descriptions of two Corals from Madeira, Primnoa and Mopsea. *Loc. cit.* pp. 245–246.

1863. Ibid.—Description of a new species of Juncella. *Loc. cit.* pp. 505–506.

1864. Ibid.—Description of a new species of Flexible Coral belonging to the genus Juncella obtained at Madeira (J. flagellum). Ann. Mag. Nat. Hist. 3 ser. xiv. p. 142.

1886. Jungersen, H. F. E.—Kara-Havets Alcyonider. Kjöbenhaven, Dijmphna Togtets Zoologisk-Botanishe Udbytte, pp. 375–380, 2 pls.

1888. Ibid.—Zeitschr. f. wiss. Zool. xlvii. p. 626 (Development of Pennatula).

1891. Ibid.—Ceratocaulon wandeli en ny Nordisk Alcyonide. Naturhist. Forening Kjöbenhaven.

1904. Ibid.—The Danish Ingolf Expedition Pennatulida, v. pt. i. 3 pls. 1 chart and 3 figs. in text.

1870. Kent, W. Saville.—On two new genera of Alcyonoid Corals taken in the recent expedition of the yacht Norna off the west coast of Spain and Portugal. Quart. Journ. Micr. Sci. x. pp. 397–399, 1 pl.

1870. Ibid.—On the Calcareous Spicula of the Gorgonaceæ; their modification of form and the importance of their characters as a basis for generic and specific diagnosis. Monthly Micr. Journ. iii.

1902. Kishinouye, K.—Preliminary Note on the Corallidæ of Japan. Zool. Anzeig. xxvi. pp. 623–626.

1877. Klunzinger, C. B.—Die Korallthiere des rothen Meeres. Part i. Die Alcyonarien und Malacodermen, p. 59, 4 pls.

1858. Kner.—Ueber Virgularia multiflora, n. sp., aus der Familie der Seefedern. Verb. zool.-bot. Ges. Wien.

1874. Koch, G. von.—Anatomie der Orgelkoralle (Tubipora hemprichii). Jena.

1878. Ibid.—Das Skelett der Alcyonarien. Morph. Jahrb. iv.

1882. Ibid.—Anatomie der Clavularia prolifera, n. sp., nebst einigen vergleichenden Bemerkungen. Morph. Jahrb. vii. p. 462 ff.

1882. Ibid.—Vorlaüfige Mittheilungen über die Gorgonien von Neapel und über die Entwickelung der Gorgonia verrucosa. MT. Zool. Stat. Neapel, iii. p. 537 ff.

1886. Koch, G. von.—Neue Anthozoen aus dem Golf von Guinea. Marburg (Itephritus = Bellonella).

1890. Ibid.—Die Alcyonaceen des Golfes von Neapel. MT. Zool. Stat. Neapel, ix. Heft 4.

1887. Ibid.—Fauna und Flora des Golfes von Neapel. Die Gorgoniden, xv. p. 99, 10 pls.

1891. Ibid.—Die systematische Stellung von Sympodium coralloides. Zool. Jahrb. v. p. 76.

1865. Kölliker, A.—Icones Histiologicæ, pp. 131–142, 8 pls.

1870. Ibid.—Beiträge zur Kenntniss der Polypen. (1) Ueber der Gattung Solanderia, pp. 11–16; (2) Semperina rubra, eine neue Gattung, pp. 17–20; (3) Ueber eine neue Alcyonarien Pseudogorgia godeffroyi. Verh. Phys-Med. Ges. Würzburg, N. F. ii. p. 22, 2 pls.; Sitzber. für 1870, pp. vii-viii.

1870. Ibid.—Anatomische systematische Beschreibung der Alcyonarien. Abth. i.

1872. Ibid.—Die Pennatuliden. Abh. Senckenberg. Nat. Ges. vii. and viii. p. 458, pls. 24.

1874. Ibid.—Die Pennatuliden Umbellula und zwei neue Typen der Alcyonarien. Festschrift Phys-Med. Ges. Würzburg, pp. 6–23, 2 pls.

1875. Ibid.—Ueber den Bau und der systematische Stellung der Gattung Umbellularia. Verb. Phys-Med. Ges. Würzburg, N. F. viii. pp. 13–18.

1880. Ibid.—Pennatulida of "Challenger," vol. i.

1848. Koren and Danielssen.—Nyt. Mag. Naturvidensk.

1858. Ibid.—Forhandl. Vidensk. Selsk. Christiania.

1877. Ibid.—Bidrag til de ved den norske kyst levende Pennatuliders Naturhistorie. Fauna littoralis norvegicæ, 3 Heft, pp. 82–103.

1883. Ibid.—Bergen's Museum Nye Alcyonider, Gorgonider, og Pennatulider tilhorende Norges Fauna. Bergen, p. 38, 13 pls.

1884. Ibid.—The Norwegian North-Atlantic Expedition, 1876–1878, Zoology—Pennatulida, p. 70. 13 pls. and 1 map.

1873. Kowalevsky, A. O.—Untersuchungen über die Entwickelung der Coelenteraten. Nachr. Ges. Freunde Naturk. Anthrop. Ethnogr. Moskau. x. (Russian).

1879. Ibid.—Zur Entwickelungsgeschichte der Alcyoniden Sympodium coralloides und Clavularia crassa. Zool. Anzeig. ii. pp. 491–493.

1883. Kowalevsky et Marion.—Documents pour l'histoire embryogénique des Alcyonaires. Ann. Mus. Marseille-Zool. i. Mém. 4.

1881. Krukenberg, C. Fr. W.—Vergleichend-physiologische Studien. Abth. v. Heidelberg; also Abth. i. 1880.

1895. Kükenthal, W.—Alcyonaceen von Ternate. Zool. Anzeig. Nos. 488 and 489.

1896. Ibid.—Alcyonaceen von Ternate. Abh. Senakenberg. Nat. Ges. xxiii. Heft 1, 4 pls.

1902. Ibid.—Diagnosen neuer Alcyonarien aus der Ausbeute der Deutschen Tiefsee Expedition Zool. Anzeig. xxv. pp. 229–303.

1902. Ibid.—Versuch einer Revision der Alcyonarien. I. Die Familie der Xeniiden. Zool. Jahrb. xv. pp. 635–662.

1903. Ibid.—Versuch einer Revision der Alcyonarien. II. Die Familie der Nephthyiden, 1 Teil. Zool. Jahrb. xix. pp. 99–172.

1904. Ibid.—Ueber einige Korallenthiere des Roten Meeres. Festschrift von E. Haeckel, pp. 33–58, 2 pls.

1904. Ibid.—Ueber eine neue Nephthyidengattung aus dem südatlantischen Ocean. Zool. Anzeig. xxvi. pp. 272–275.

1904. Ibid.—Diagnosen neuer Alcyonarien aus der Ausbeute der Deutschen Tiefsee Expedition. Zool. Anzeig. xxv. pp. 299–303, also p. 596 (Umbellula).

1905. Ibid.—Versuch einer Revision der Alcyonarien. II. Die Familie der Nephthyiden, 2 Teil. Zool. Jahrb. xxxi. pp. 503–726, 7 pls. and 61 figs.

1864. Lacaze Duthiers.—Histoire naturelle des Corail. Paris.

1864. Lacaze-Duthiers.—Couleur der Alcyonaires. Comptes Rendus, lix. p. 252.

1865. Ibid.—Des Sexes chez Alcyonaires. Comptes Rendus, lx. pp. 840–843.

1865. Ibid.—Note sur la présence dos Kophobelemnon dans les eaux de Banyuls. Comptes Rendus, cxii. pp. 1294–1297.

1900. Ibid.—Coralliaires du Golfo du Lion, Alcyonaires. Arch. Zool. Exp. (3) viii. pp. 353–462, 5 pls.

1816. Lamarck.—Histoire naturelle des Animaux sans Vertèbres. Vol. ii. Paris.

1812. Lamouroux.—Genus Telesto. Bull. Soc. Philom. Paris.

1816. Ibid.—Histoire des Polypiers coralligènes flexibles. Caen.

1821. Ibid.—Exposition methodique des genres de l'ordre des Polypiers. Paris.

1874. Lindahl, J.—Umbellula from Greenland. Ann. Mag. Nat. Hist. 4 ser. xiii. p. 258.

1875. Ibid.—Sur les Ombellules. Journ. d. Zool. iv. pp. 440–442, 1 pl.

1874. Ibid.—Om Pennatulid-Slägtet Umbellula (Cuv.). Svenska Vet. Akad. Handl. xiii. No. 3, 3 pls.

1758. Linné.—Systema Naturæ. T. i. ed. x.

1877. Marenzeller.—Die Cœlenteraten, Echinodermen und Würmer der K. K. Oesterr.-Ung. Nordpol-Exp. Wien.

1886. Ibid.—Ueber die Sarcophytum benannten Alcyoniden. Zool. Jahrb. i. pp. 341–368, 1 pl.

1886. Ibid.—Denkschr. Akad. Wiss. Wien, xxxv.

1884. Marshall, A. Milnes.—Report on the Pennatulida dredged by H.M.S. "Triton." Trans. Roy. Soc. Edinburgh, xxxii. pt. 1, p. 140.

1882. Marshall and Marshall.—Report on the Oban Pennatulida, p. 77, pls. 4.

1887. Marshall and Fowler.—Report on the Pennatulida of the Mergui Archipelago. Journ. Linn. Soc. London, xxi. pp. 267–286.

1887. Ibid.—Report on Pennatulida dredged by H.M.S. "Porcupine." Trans. Roy. Soc. Edinburgh, xxiii. p. 453.

1898. May, W.—Alcyonaceen von Ostspitzbergen. Nach der Ausbeute Prof. Willy Kükenthal's und Dr. Alfred Walter's im Jahr. 1889. Zool. Jahrb. xi.

1899. Ibid.—Beiträge zur Systematik und Chorologie der Alcyonaceen. Jenaische Zeitschr. Naturwiss. xxiii. p. 190, 5 pls.

1905. Ibid.—Fauna Arctica.

1905. Menneking, F.—Ueber die Anordnung der Schuppen und das Kanalsystem bei Stachyodes ambigua (Stud.), Caligorgia flabellum (Ehrbg.), Calyptrophora agassizii (Stud.), Amphilaphis abietina (Stud.), und Thouarella variabilis (Stud.). Arch. Naturges. lxxi. pp. 245–266, 2 pls.

1835. Milne-Edwards.—Mémoire sur un nouveau genre de la famille des Alcyoniens. Ann. Sci. Nat. ser. 2, iv.

1838. Ibid.—Recherches anatomiques, etc., Alcyonides, Alcyons proprement dits. Ann. Sci. Nat. 2 ser. iv. 5 pls.

1857. Milne-Edwards and Haime.—Histoire naturelle des Coralliaires ou Polypes proprement dits. 3 vols. 1857–1860.

1861. Möbius, K.—Neue Gorgoniden des naturhist. Museums zu Hamburg. Nov. act. Acad. Leop.-Carol. xxix.; also separately, Jena, 1861, p. 12, 3 pls.

1874. Ibid.—Ueber eine Hornkoralle Verucella guadelupensis. Schr. naturwiss. Ver. Schleswig-Holstein, Bd. i. Heft 2, pp. 204–206.

1902. Moroff, Th.—Einige neue Pennatuliden aus der Münchener Sammlung. Zool. Anzeig. xxv. pp. 579–582.

Ibid.—Studien über Octocorallien. Zool. Jahrb. xvii. pp. 363–410, pls. xiv.–xviii.

Ibid.—Einige neue Japanische Gorgonaceen in der Münchener Sammlung. Zool. Anzeig. xxv. pp. 582–584.

1876. Moseley, H. N.—On the structure and relations of the Alcyonarian Heliopora cœrulea. Phil. Trans. clxvi. pt. i. p. 91.

1881. Moseley, H. N.—Report on certain Hydroid, Alcyonarian, and Madreporarian Corals procured during the voyage of H.M.S. "Challenger." 1873–1876, "Challenger" Reports, vol. ii. (Heliopora, Sarcophytum.)

1873. Moss, Ed. L.—Description of a Virgularian Actinozoon (Osteocella). Proc. Zool. Soc. London, pp. 730–732.

1787. Müller, O. F.—Zoologia Danica.

1766. Pallas.—Elenchus Zoophytorum, p. 434.

1889. Pfeffer, G.—Zur Fauna von Süd-Georgien. Jahrb. Hamburg Wiss. Anst. Jahrg. vi. 2 Hälfte, 49.

1842. Philippi.—Arch. Naturges. viii.

1867. Pourtales, L. F.—Contributions to the Fauna of the Gulf Stream at great depths. 1st and 2nd series. Bull. Mus. Comp. Zool. Harvard, vol. i. pp. 103–120, 121–142.

1903. Pratt, E. M.—The Alcyonaria of the Maldives. Fauna and Geography of the Maldives and Laccadive Archipelagos, vol. ii. pt. i. pp. 503–535, pls. xxviii.–xxxi.

1905. Ibid.—The digestive organs of the Alcyonaria and their relation to the Mesoglœal Cell-plexus. Quart. Jour. Micr. Sci. xlix. pp. 327–362, 3 pls.

1905. Ibid.—The Mesoglœal Cells of Alcyonium (Preliminary Account). Zool. Anzeig. xxv. pp. 545–548, 3 figs.

1905. Ibid.—On some Alcyonidæ. Ceylon Pearl Oyster Report, Supp. Rep. xix. pp. 247–268, 3 pls.

1902. Pruvot, G.—Review of May: La Faune des Alcyonaires arctiques, subarctiques et antarctiques. Fauna Arctica, i. pp. 379–408.

1900. Pütter, H. — Alcyonaceen des Breslauer Museums. Zool. Jahrb., Band xiii. Heft 5, 2 pls.

1834. Quoy et Gaimard.—Voyage de l'Astrolabe. Zoophytes iv. Paris.

1869. Richiardi.—Monograffia della Famiglia dei Pennatularii. Bologna.

1882. Ridley, Stuart O.—Contributions to the knowledge of the Alcyonaria. Part i. Ann. Mag. Nat. Hist. 5 ser. ix. pp. 125–133, 1 pl.

1882. Ibid.—Part ii. *Loc. cit.* 5 ser. ix. pp. 184–193.

1882. Ibid.—Proc. Zool. Soc. London, p. 231, etc.

1883. Ibid.—The Coral-fauna of Ceylon, with descriptions of new species. Ann. Mag. Nat. Hist. 5 ser. xi. pp. 250–262.

1887. Ibid.—Report on the Alcyoniid and Gorgoniid Alcyonaria of the Mergui Archipelago, collected for the Indian Museum. Jour. Linn. Soc. London, xxi. pp. 223–247, 2 pls.

1884. Ibid.—Zoological Collections of H.M.S. "Alert." London.

1896. Roule, L.—Resultats scientifique de la Campagne du "Caudan" dans le golfe de Gascogne. Ann. de l'Université de Lyon, xxvi.

1902. Ibid.—Anthozoa Alcyonaria (Clavularia), "Southern Cross" Collection. London, pp. 290–293, pl. xlvii. figs. 1–3.

1846. Sars.—Fauna Littoralis Norvegiæ, i.

1817. Savigny.—Description de l'Égypte. Hist. Nat. Suppl. i., Paris.

1896. Schenk, A.—Clavulariiden, Xeniiden und Alcyoniiden von Ternate. Abh. Senckenberg. Nat. Ges. xxiii. Heft 1, pp. 41–80, 3 pls.

1905. Simpson, Jas. J.—A New Cavernularid from Ceylon (Fusticularia). Ann. Mag. Nat. Hist. 7 ser. xv. pp. 561–565, 1 pl.

1905. Ibid.—Agaricoides, a new type of Siphonogorgid Alcyonarian. Zool. Anzeig. xxix. pp. 263–271, 19 figs.

1906. Ibid.—The Structure of Isis hippuris Linnæus. Jour. Linn. Soc. Zoology, London, vol. xxxvii. pp. 421–434, 1 pl.

1873. Stearne, R. S. C.—Description of a new species of Alcyonoid Polyp (Pavonaria blakei). San Francisco Mining and Scientific Press.

1874. Ibid.—Description of a new genus and species of Alcyonoid Polyp (Verrilliablakei). Proc. Calif. Acad. Sci. v. pp. 147–149. Amer. Jour. Sc. 3 ser. vii. pp. 68–70.

1874. Ibid.—Remarks on a new Alcyonoid Polyp from Burrard's Inlet. Proc. Calif. Acad. Sci. v. pp. 7–12.

1874. Ibid.—Verrillia blakei or Halipteris blakei. Amer. Naturalist, xvi. p. 55.

1855. Stimpson.—Description of some of the marine Invertebrates in the Chinese and Japanese Seas. Proc. Acad. Nat. Sci. Philadelphia, vii.

1873. Studer, Th.—Ueber Bau und Entwicklung des Achse von Gorgonia bartholomi. MT. nat. Ges. Bern. pp. 85–97, 3 pls.

1875. Ibid.—Solenocaulon. Monatsber. Preuss. Akad. Wiss. Berlin, p. 668.

1878. Ibid.—Uebersicht der Anthozoa Alcyonaria welche während der Reise S. M. S. "Gazelle" um die Erde gesammelt wurden. Monatsber. Akad. Wiss. Berlin, pp. 632–688, 5 pls.

1887. Ibid.—Versuch eines Systems der Alcyonaria. Arch. Naturges. liii. pp. 1–74, 1 pl.

1888. Ibid.—On some new species of the Genus Spongodes from the Philippine Is. and the Japanese Seas. Ann. Mag. Nat. Hist. pp. 69–72.

1889. Ibid.—Supplementary Report on Alcyonaria of "Challenger," vol. xxxii.

1891. Ibid.—Cas de Fissiparite chez un Alcyonaire. Bull. Soc. Zool. France, xvi. Paris.

1894. Ibid.—Note preliminaire sur les Alcyonaires: Report on Dredging operations of Steamer "Albatross." Bull. Mus. Comp. Zool. Harvard, xxv. pp. 53–69.

1894. Ibid.—Alcyonarien aus der Sammlung des Naturhist. Museums in Lübeck. MT. Geogr. Ges. und des Naturhist. Museums. Lübeck, 2 ser. Heft 7 u. 8.

1901. Ibid.—Alcyonaires provenant des Campagnes de l'Hirondelle: Resultats des Campagnes scientifiques du Prince de Monaco. Fasc. xx. p. 64, 11 pls.

1872. Targioni-Tozzetti, A.—Nota intorno ad alcuni forme di Alcionari e di Gorgonacei, etc. Atti Soc. Ital. Sci. Nat. Mus. Milano, xv. pp. 453–459.

1841. Templeton.—Descriptions of a few Invertebrated Animals obtained at the Isle de France. Trans. Zool. Soc. London, vol. ii.

1905. Thomson, J. Arthur.—Ceylon Pearl Oyster Report; Alcyonaria. Supplementary Report, xxvii. pp. 167–186, 1 pl.

1905. Thomson, J. A., and Henderson, W. D.—Alcyonaria; Ceylon Pearl Oyster Fisheries Report. Royal Society, London, pp. 271–328, 6 pls.

1905. Ibid.—Preliminary Notice of the Deep-Sea Alcyonaria collected in the Indian Ocean. Ann. Mag. Nat. Hist., 7 ser. xv. pp. 547–557.

1905. Thomson, J. A., and Ritchie, J.—The Alcyonarians of the Scottish National Antarctic Expedition. Trans. Roy. Soc. Edinburgh, xli. pp. 851–860, 2 pls.

1890. Thurston, E.—Notes on the Pearl, etc., Fisheries of the Gulf of Manaar. Government Central Museum, Madras.

Valenciennes.—MSS. coll. du Mus. Jardin des Plantes, Paris.

1855. Ibid.—Extrait d'une monographie de la famille des Gorgonides de la classe des Polypes. Comptes Rendus, xli.

1864. Verrill, A. E.—Notice of a Primnoa from St. George's Bank. Proc. Essex Inst. iii. pp. 127–129.

1864. Ibid.—List of the Polyps and Corals (sent by the Museum of Comparative Zoology to other institutions in exchange), with annotations. Bull. Mus. Comp. Zool. Harvard, i. No. 3, p. 29.

1866. Ibid.—Classification of Polyps. (Proc.) Communicat. Essex Instit. iv. Salem.

1866. Ibid.—Lissogorgia, n. sp. Proc. Boston Soc. Nat. Hist. x. pp. 22, 23.

1866. Ibid.—On the Polyps and Corals of Panama. Proc. Boston Soc. Nat. Hist. x. pp. 323–333.

1866. VERRILL, A. E.—Synopsis of Polyps and Corals of N. Pacific Expedition. Part ii. Alcyonaria. (Proc.) Communicat. Essex Instit. vol. iv.

1866. IBID.—Revision of the Polypi of Eastern Coast of U.S. America. Mem. Boston Soc. Nat. Hist. i. pp. 1–45.

1868. IBID.—Critical Remarks on the Halcyonoid Polyps in the Museum of Yale College, with descriptions of new Genera. Amer. Journ. Sci. 2 ser. xlv. pp. 411–416.

1868. IBID.—*Loc. cit.* 2 ser. xlvi. pp. 143, 144.

1869. IBID.—*Loc. cit.* 2 ser. xlvii. p. 286.

1869. IBID.—*Loc. cit.* 2 ser. xlviii. pp. 419, 420.

1870. IBID.—*Loc. cit.* 2 ser. xlix. pp. 370–375.

1872. IBID.—*Loc. cit.* 3 ser. iii.

1874. IBID.—*Loc. cit.* 3 ser. vii.

1878. IBID.—Notice of recent additions to the marine fauna of the Eastern Coast of N. America. *Loc. cit.* 3 ser. xvi.

1879. IBID.—*Loc. cit.* 3 ser. xvii.

1882. IBID.—Notice of the remarkable marine fauna occupying the outer banks of the Southern Coast of New England. *Loc. cit.* xxiii.

1878. IBID.—Description of a new species of Paragorgia (pacifica). Canad. Naturalist, viii. p. 476.

1883. IBID.—Rep. U.S. Comm. of Fish and Fisheries.

1883. IBID.—Report on the Anthozoa and on some additional species dredged by the "Blake" in 1877–1879, and by the U.S. Fish Commission Steamer "Fish-Hawk" in 1880–1882. Bull. Mus. Comp. Zool. Harvard, xi. No. 1.

1902. IBID.—Trans. Connect. Acad. xi. (Eunicea atra, Verrucella grandis).

1902. VERSLUYS, J.—Siboga Expeditie Monographie, xiii. Die Gorgoniden der Siboga Expedition. I. Chrysogorgiidæ, pp. 117, 170 figs.

1906. IBID.—Jets over Zoogeographie, naar aarleidung von de Marine Fauna von den oost-indischen Archipel. Handb. Nederland Nat. Congr. Aarnhem, pp. 490–511.

1906. IBID.—Siboga-Expeditie Monographie, xiii. (*a*).. Die Gorgoniden der Siboga-Expedition, ii. Die Primnoidæ, p. 183, 10 pls. 178 figs. and 1 map.

1888. VIGUIER, C.—Études sur les animaux inférieurs de la baie d'Alger. Arch. Zool. Exper. 2 ser. vi. pp. 351–373, 2 pls. (Fascicularia)

1897. WHITELEGGE, TH.—The Alcyonaria of Funafuti, pt. ii. Mem. Australian Mus. iii. pt. 5, Nov. 17, pp. 307–320, 2 pls.

1883. WILSON, E. B.—The development of Renilla. Phil. Trans. clxxiv. p. 723 et seq.

1884. IBID.—The Mesenterial Filaments of the Alcyonaria. MT. Zool. Stat. Neapel, v.

1902. IBID.—Notes on Merogony and regeneration in Renilla. Biol. Bull. iv. pp. 215–226, 4 figs.

1864. WRIGHT, E. P.—On the new genus of Alcyonidæ (Hartea elegans). Proc. Dublin Microscop. Club, p. 6, 1 pl.

1865. IBID.—On a new genus of Alcyonidæ (Hartea elegans). Quart. Journ. Micr. Sci. v. pp. 213–217.

1869. IBID.—On a new genus of Gorgonidæ (Keratoisis). Ann. Mag. Nat. Hist. 4 ser. iii. pp. 23–27.

1869. IBID.—Notes on the Animal of the Organpipe Coral (Tubipora musica). Ann. Mag. Nat. Hist. 4 ser. iii.

1886. IBID.—On a new genus of Alcyonidæ, 1 pl.

1889. WRIGHT, E. P., and STUDER, TH.—Report on the Scientific Results of the voyage of H.M.S. "Challenger." Alcyonaria, vol. xxxi.

1889. IBID.—Arch. Naturges. liii.

O

EXPLANATION OF PLATES.

PLATE I.

FIG. 1. *Sarcophytum aberrans*, n. sp. STALKED COLONY without basal attachment. Slightly enlarged (× 1½).

„ 2. *Sarcophytum aberrans*, n. sp. ENCRUSTING COLONY, (*a*) portion enlarged (× 1⅓), showing part of the spiral bare streak, at one place the cœnenchyma overarching it, at another the cœnenchyma completely surrounding the support; (*b*) distal end of an autozooid showing the arrangement of the pinnules on the tentacles; (*c*) colony reduced (⅓ n. s.) in size showing the incrusting habit and the projecting siliceous support.

„ 3. *Sarcophytum agaricoides*, n. sp. (Nat. size.) Note the characteristic autozooids; the siphonozooids (represented by white spots) are much more numerous than in the figure.

„ 4. *Spongodes alcocki*, n. sp. (× 15). Two polyps showing the single projecting spicule of the Stützbündel.

„ 5. *Paragorgia splendens*, n. sp. (Nat. size.) Note the characteristic clusters of autozooids.

„ 6. *Keroëides koreni*, Wright and Studer (× 10). Small portion of a twig.

„ 7. *Keroëides koreni*, Wright and Studer. Colony nat. size.

„ 8. *Muricella bengalensis*, n. sp. Small portion enlarged (× 20). Note the large size of the spicules of the cœnenchyma.

„ 9. *Pleurocorallium variabile*, n. sp. Small portion enlarged (× 8) showing the arenaceous cœnenchyma and the stellate appearance of the retracted polyps.

„ 10. *Juncella elongata*, Pallas. Small portion enlarged (× 4).

PLATE II.

FIG. 1. *Stachyodes allmani*, Wright and Studer. Axis of a colony to show mode of branching. (Nat. size.)

„ 2. *Acanthogorgia aspera*, Pourtales. Colony nat. size.

„ 3. *Chrysogorgia flexilis*, Wright and Studer. Colony showing branching and arrangement of polyps.

„ 4. *Chrysogorgia irregularis*, n. sp. Part of a colony to show the indefinite mode of branching.

„ 5. (*a* and *b*) *Stachyodes allmani*, Wright and Studer. Obverse and reverse views of a small portion of a twig enlarged.

„ 6. *Sympodium indicum*, n. sp. Portion enlarged showing burrowing polychæt worm.

„ 7. *Sympodium incrustans*, n. sp. Part enlarged to show the furrowing of the polyps and also stages in retraction.

PLATE III.

FIG. 1. *Protoptilum medium*, n. sp. Part enlarged. Note the irregular margins of the calyces.

„ 2. *Thouarella moseleyi*, var. *spicata*, n. Portion of colony nat. size showing the mode of branching.

„ 3. *Paramuricea indica*, n. sp. Part enlarged to show the markedly spinose character of the cœnenchyma.

Fig. 4. *Thouarella moseleyi*, var. *spicata*, n. Portion enlarged (× 20). Note the spines on the pre-opercular scales.

„ 5. *Lepidogorgia verrilli*, Wright and Studer. (*a*) Small portion of colony nat. size; (*b*) part enlarged (× 15) showing the dense armature and a partially retracted polyp.

„ 6. *Chrysogorgia indica*, n. sp. (× 1½). To show mode of branching and arrangement of polyps.

„ 7. *Acamptogorgia bebrycoides*, var. *robusta*, n. Fragment nat. size.

„ 8. *Acamptogorgia bebrycoides*, var. *robusta*, n. Twig enlarged (× 10). Note the short and cylindrical polyps and the almost horizontal operculum.

Plate IV.

Fig. 1. *Keroëides gracilis*, Whitelegge. Portion of colony nat. size.

„ 2. *Keroëides gracilis*, Whitelegge. Part of twig enlarged (× 10). Note the large and regularly arranged spicules.

„ 3. *Keroëides gracilis*, Whitelegge. Portion of main stem enlarged (× 10) showing the smaller and more indefinitely arranged spicules.

„ 4. *Parisis indica*, n. sp. Branch nat. size.

„ 5. *Parisis indica*, n. sp. Part of axis enlarged showing the clusters of epizoic *Palythoa*.

„ 6. *Sympodium tenue*, n. sp. Colony (× 1½) showing the support of siliceous sponge spicules.

„ 7. *Distichoptilum gracile*, Verrill. Part enlarged (× 10). Note the small scarcely projecting polyps.

„ 8. *Parisis indica*, n. sp. Basal portion (× 1½) showing the disc of attachment.

„ 9. *Parisis indica*, n. sp. Part of a branch enlarged (× 1½). Contrast the proportions of node and internode with those in fig. 8.

„ 10. *Chironephthya macrospiculata*, n. sp. Terminal portion enlarged (× 12). Note the enormous spindles of the cœnenchyma.

„ 11. *Suberogorgia köllikeri*, var. *ceylonensis*, Thomson. Branch nat. size.

„ 12. *Suberogorgia köllikeri*, var. *ceylonensis*, Thomson. Portion enlarged (× 10).

Plate V.

Fig. 1. *Acanthomuricea ramosa*, n. g. et sp. Terminal twig enlarged (× 10). Side to which the polyps are directed.

„ 2. *Stereacanthia indica*, n. g. et sp. Part enlarged (× 10).

„ 3. *Acamptogorgia circium*, n. sp. Twig (× 2). Note the spinose appearance.

„ 4. *Acanthomuricea ramosa*, n. g. et sp. Reverse side of fig. 1.

„ 5. *Umbellula rosea*, n. sp. Rachis nat. size.

„ 6. *Pleurocorallium variabile*, n. sp. Portion enlarged (× 12). Note the arenaceous appearance of the cœnenchyma and the stellate form of the polyps.

„ 7. *Juncella miniacea*, n. sp. Colony nat. size. Note gall.

„ 8. *Acanthomuricea ramosa*, n. g. et sp. Branch (× 1½).

„ 9. *Paragorgia splendens*, n. sp. Branch nat. size.

„ 10. *Astrogorgia rubra*, n. sp. Terminal portion of twig (× 10).

„ 11. *Acis spinosa*, n. sp. Part of twig enlarged (× 14). Note the difference in the spiculation of contiguous portions; also the enormous plate-like spicules.

„ 12. *Juncella miniacea*, n. sp. Part enlarged (× 12) showing the characteristic polyps.

„ 13. *Stenella horrida*, n. sp. Terminal twig enlarged (× 12). Note the rugose appearance.

„ 14. *Paragorgia splendens*, n. sp. Cluster of autozooids enlarged; also showing the arrangement of the siphonozooids.

„ 15. *Acanthogorgia aspera*, Pourtales. Side and end view of polyps (× 10).

Plate VI.

Fig. 1. *Thesioides inermis*, n. g. et sp. Part of colony nat. size.
„ 2. *Thesioides inermis*, n. g. et sp. Portion enlarged (× 6). Note the peculiar tentacles.
„ 3. *Chrysogorgia dichotoma*, n. sp. Axis nat. size to show mode of branching.
„ 4. *Acamptogorgia bebrycoides*, von Koch. Branch (× 2).
„ 5. *Acamptogorgia bebrycoides*, von Koch. Terminal twig enlarged (× 10).
„ 6. *Ceratoisis gracilis*, n. sp. Part enlarged (× 11) to show characteristic armature of polyps; (*a*) small portion nat. size.
„ 7. *Sympodium pulchrum*, n. sp. Part of colony nat. size encrusting a sponge skeleton.
„ 8. *Sclerobelemnon kölliкeri*, n. sp. Portion enlarged (× 5½). Note the peculiar pseudo-calyx and the siphonozooids.

Plate VII.

Fig. 1. *Pennatula pendula*, n. sp. A pinnule. Note the rugose autozooids and the ova appearing through the semi-transparent walls.
„ 2. *Chrysogorgia orientalis*, Versluys. Portion enlarged to show the mode of branching and the characteristic architecture of the polyps.
„ 3. *Protocaulon indicum*, n. sp. Part enlarged. Note the long polyps and the peculiar flagelliform tentacles.
„ 4. *Bathyptilum indicum*, n. sp. Colony nat. size.
„ 5. *Stachyptilum maculatum*, n. sp. Colony nat. size.
„ 6. *Umbellula elongata*, n. sp. Portion of autozooid cut open to show the embryos.
„ 7. *Protocaulon indicum*, n. sp. Colony nat. size.
„ 8. *Anthoptilum decipiens*, n. sp. Part enlarged showing the reproductive bodies appearing through the thin body-walls.
„ 9. *Stachyptilum maculatum*, n. sp. Part enlarged showing the strong autozooids and the conspicuous stellate siphonozooids.
„ 10. *Funiculina gracilis*, n. sp. Portion enlarged to show the arrangement of the autozooids and prominent projecting eight-lobed siphonozooids in an almost definite row on the metarachidial surface.

Plate VIII.

Fig. 1. *Pennatula indica*, n. sp. (*a*) Complete colony nat. size; (*b*) portion of rachis enlarged (× 10); (*c*) part of a pinnule enlarged (× 10).
„ 2. *Muricella bengalensis*, n. sp. Twig enlarged (× 8).
„ 3. *Umbellula purpurea*, n. sp. Terminal portion nat. size.
„ 4. *Agaricoides alcocki*, Simpson, n. g. et sp. Colony nat. size.
„ 5. *Pennatula splendens*, n. sp. Portion of rachis with four pinnules (× 4).
„ 6. *Spongodes alcocki*, n. sp. Terminal portion (× 10).
„ 7. *Pennatula pendula*, n. sp. Colony nat. size.
„ 8. *Pennatula veneris*, n. sp. Pinnule (× 6).
„ 9. *Umbellula dura*, n. sp. Colony nat. size.
„ 10. *Pennatula pendula*, n. sp. Pinnule (× 6).

Plate IX.

Fig. 1. *Acamptogorgia circium*, n. sp.
„ 2. *Distichoptilum gracile*, Verrill.
„ 3. *Stenella horrida*, n. sp.

FIG. 4. *Funiculina gracilis*, n. sp.
„ 5. *Acanthomuricea ramosa*, n. g. et sp.
„ 6. *Chrysogorgia irregularis*, n. sp.
„ 7. *Sarcophytum aberrans*, n. sp.
„ 8. *Sympodium decipiens*, n. sp.
„ 9. *Calicogorgia rubrotincta*, n. g. et sp.
„ 10. *Calicogorgia investigatoris*, n. g. et sp.
„ 11. *Sarcophytum aberrans*, n. sp. (more highly magnified).
„ 12. *Acanthomuricea spicata*, n. g. et sp.
„ 13. *Pleurocorallium variabile*, n. sp.
„ 14. *Acanella rigida*, Wright and Studer.
„ 15. *Scirpearella alba*, n. sp.
„ 16. *Stachyptilum maculatum*, n. sp.
„ 17. *Juncella elongata*, Pallas.
„ 18. *Sympodium indicum*, n. sp.
„ 19. *Stereacanthia indica*, n. g. et sp.
„ 20. *Paramuricea indica*, n. sp.

PLATE X.

Agaricoides alcocki, Simpson, n. g. et sp.

FIG. 1. Typical colony nat. size.
„ 2. Zooid enlarged (× 10).
„ 3. Young colony (nat. size) showing first stage in the development of the verrucæ.
„ 4. Enlarged polyp and stem canal showing the method of growth.
„ 5. Vertical section through a colony to show the relations of the parts (nat. size).
„ 6. Enlarged section of one of the stem canals (× 60).
„ 7. Asulcar filament enlarged (× 80).
„ 8. Cross section through the introverted anthocodia (× 30).
„ 9. Vertical section through a verruca and stem canal with the zooid introverted (× 10).
„ 10. Pale yellow spicules of the anthocodia.
„ 11. Enlarged section through the stomodæum showing the sulcus (× 70).
„ 12. Transparent spicules of the partition walls.
„ 13. Transparent spicules of the outer wall of the trunk.
„ 14. Transparent and pale yellow spicules of the verruca-disc.
„ 15. Spicule enlarged (× 490) to show the organic axis.
„ 16. Surface view of a verruca-disc, the zooid being introverted (× 7).
„ 17. Enlarged section through a wall showing the spicule cavities and the residue of the organic axis (× 15).
„ 18. Pale yellow spicules of the tentacles.
„ 19. Section through the trunk showing the mesenterial filaments passing into the stem canals (× 60).

PLATE I.

FIG. 1. *Sarcophytum aberrans*, n. sp. STALKED COLONY without basal attachment. Slightly enlarged (× 1½).

„ 2. *Sarcophytum aberrans*, n. sp. ENCRUSTING COLONY, (*a*) portion enlarged (× 1⅓), showing part of the spiral bare streak, at one place the cœnenchyma overarching it, at another the cœnenchyma completely surrounding the support; (*b*) distal end of an autozooid showing the arrangement of the pinnules on the tentacles; (*c*) colony reduced (⅓ n. s.) in size showing the incrusting habit and the projecting siliceous support.

„ 3. *Sarcophytum agaricoides*, n. sp. (Nat. size.) Note the characteristic autozooids; the siphonozooids (represented by white spots) are much more numerous than in the figure.

„ 4. *Spongodes alcocki*, n. sp. (× 15). Two polyps showing the single projecting spicule of the Stützbündel.

„ 5. *Paragorgia splendens*, n. sp. (Nat. size.) Note the characteristic clusters of autozooids.

„ 6. *Keroëides koreni*, Wright and Studer (× 10). Small portion of a twig.

„ 7. *Keroëides koreni*, Wright and Studer. Colony nat. size.

„ 8. *Muricella bengalensis*, n. sp. Small portion enlarged (× 20). Note the large size of the spicules of the cœnenchyma.

„ 9. *Pleurocorallium variabile*, n. sp. Small portion enlarged (× 8) showing the arenaceous cœnenchyma and the stellate appearance of the retracted polyps.

„ 10. *Juncella elongata*, Pallas. Small portion enlarged (× 4).

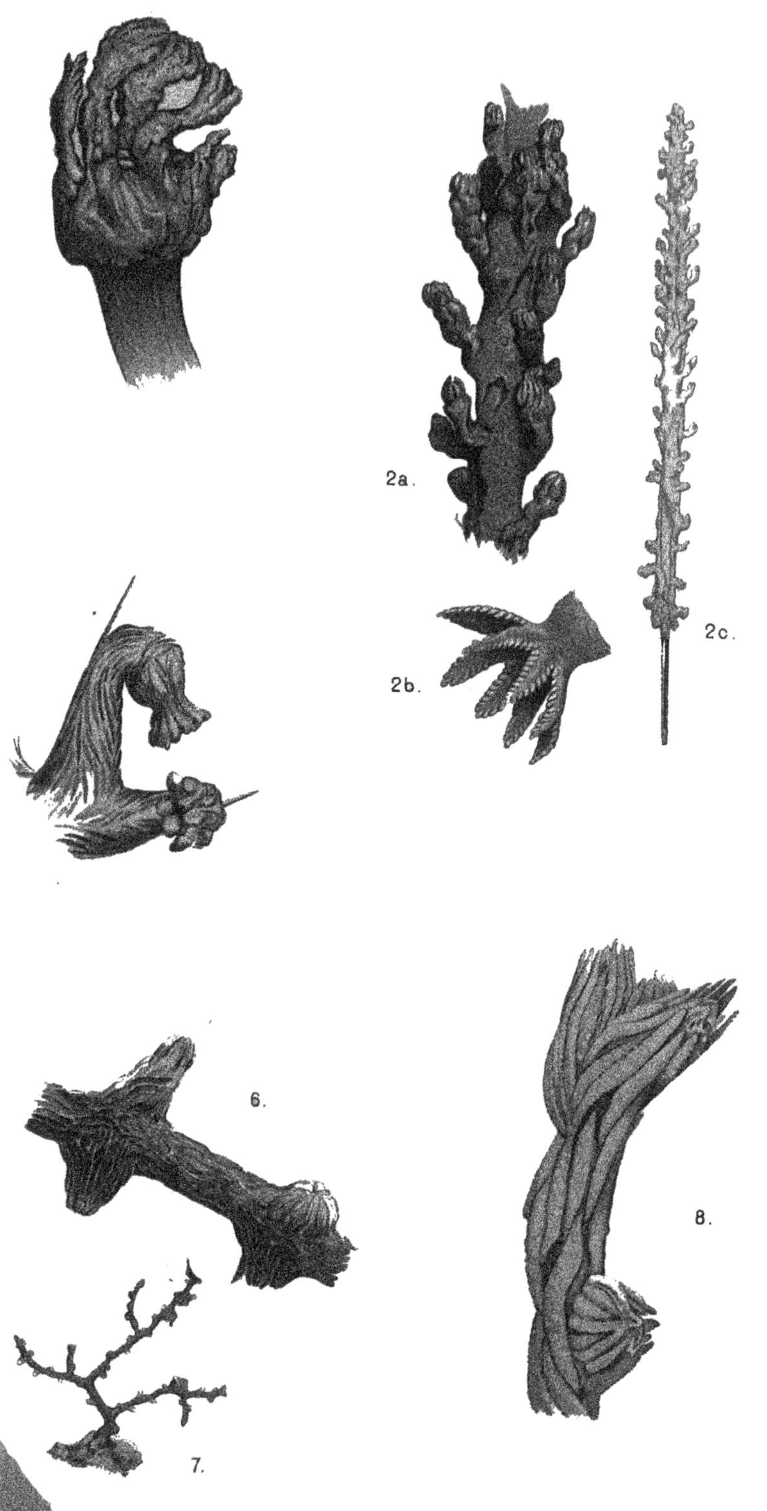
2a.
2b.
2c.
5.
6.
7.
8.

Plate II.

Fig. 1. *Stachyodes allmani*, Wright and Studer. Axis of a colony to show mode of branching. (Nat. size.)

„ 2. *Acanthogorgia aspera*, Pourtales. Colony nat. size.

„ 3. *Chrysogorgia flexilis*, Wright and Studer. Colony showing branching and arrangement of polyps.

„ 4. *Chrysogorgia irregularis*, n. sp. Part of a colony to show the indefinite mode of branching.

„ 5. (*a* and *b*) *Stachyodes allmani*, Wright and Studer. Obverse and reverse views of a small portion of a twig enlarged.

„ 6. *Sympodium indicum*, n. sp. Portion enlarged showing burrowing polychæt worm.

„ 7. *Sympodium incrustans*, n. sp. Part enlarged to show the furrowing of the polyps, and also stages in retraction.

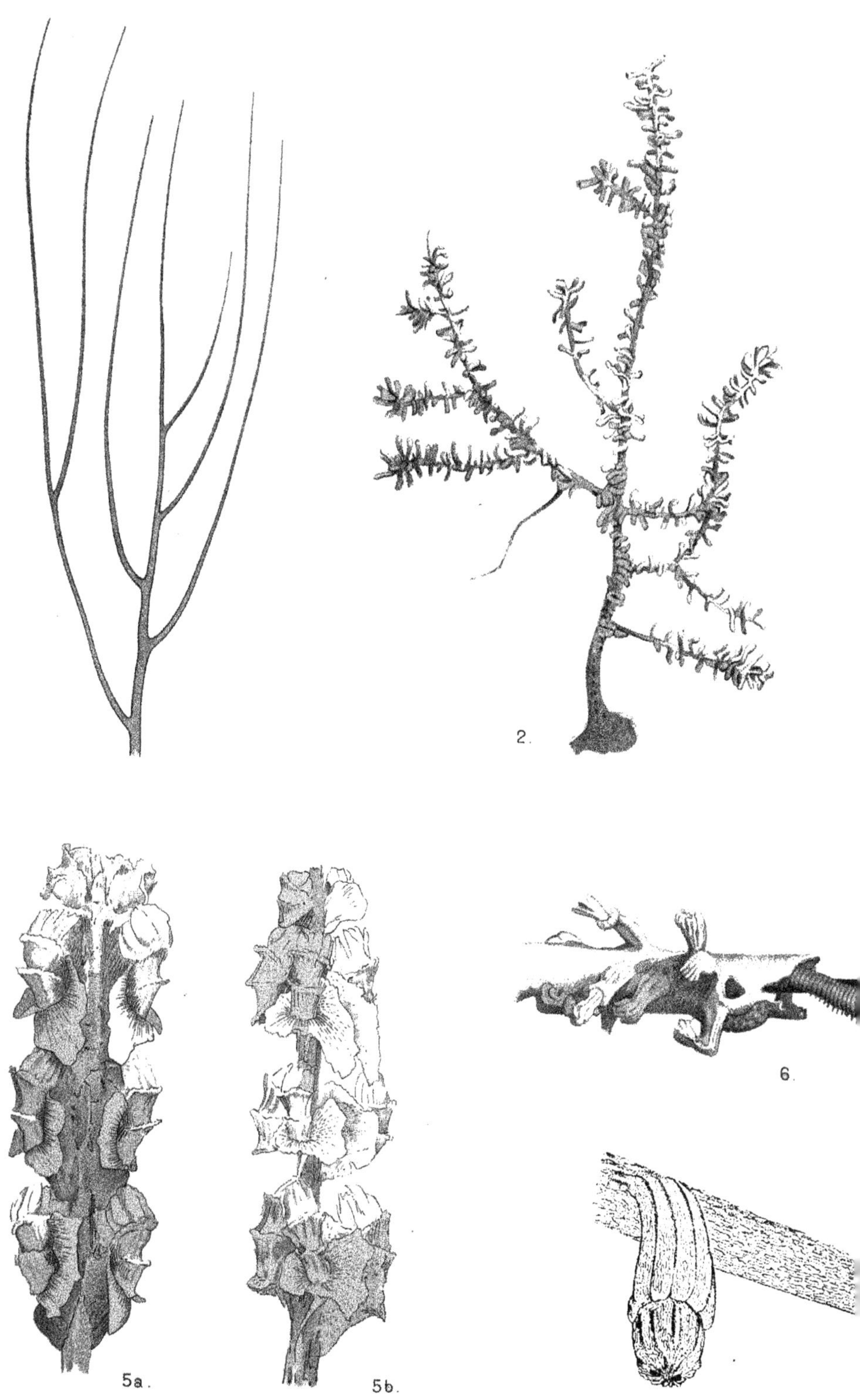
2.
5a.
5b.
6.
Davidson del.

Plate III.

Fig. 1. *Protoptilum medium*, n. sp. Part enlarged. Note the irregular margins of the calyces.

„ 2. *Thouarella moseleyi*, var. *spicata*, n. Portion of colony nat. size showing the mode of branching.

„ 3. *Paramuricea indica*, n. sp. Part enlarged to show the markedly spinose character of the cœnenchyma.

„ 4. *Thouarella moseleyi*, var. *spicata*, n. Portion enlarged (×20). Note the spines on the pre-opercular scales.

„ 5. *Lepidogorgia verrilli*, Wright and Studer. (*a*) Small portion of colony nat. size; (*b*) part enlarged (× 15) showing the dense armature and a partially retracted polyp.

„ 6. *Chrysogorgia indica*, n. sp. (× 1½). To show mode of branching and arrangement of polyps.

„ 7. *Acamptogorgia bebrycoides*, var. *robusta*, n. Fragment nat. size.

„ 8. *Acamptogorgia bebrycoides*, var. *robusta*, n. Twig enlarged (×10). Note the short and cylindrical polyps and the almost horizontal operculum.

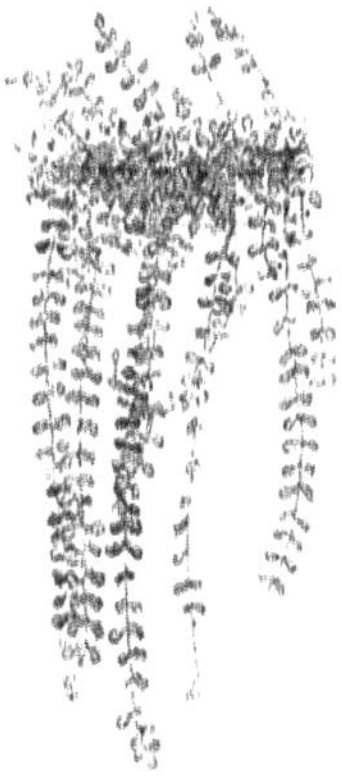

2.

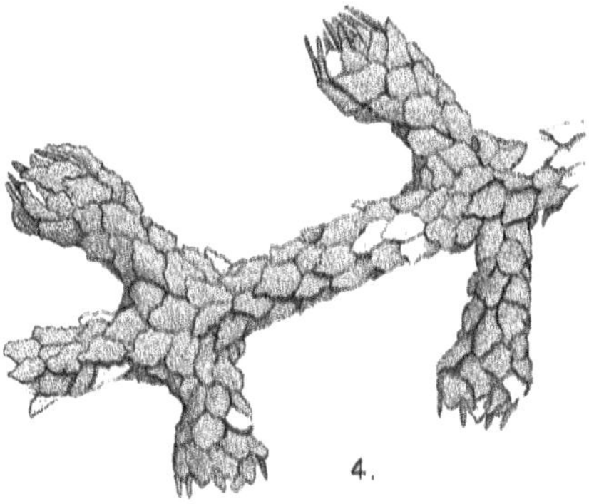

4.

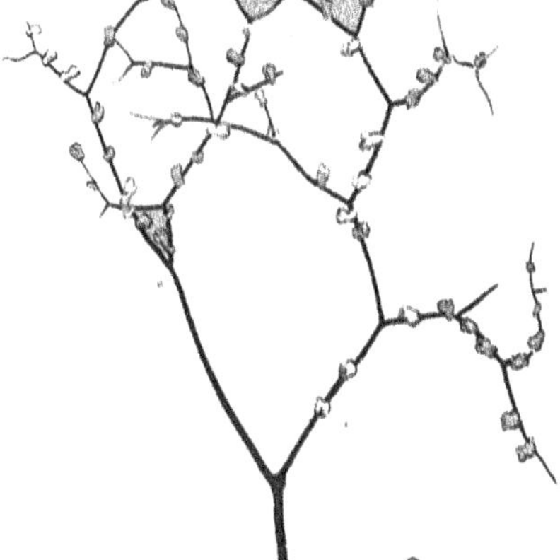

6.

PLATE IV.

FIG. 1. *Keroëides gracilis*, Whitelegge. Portion of colony nat. size.

„ 2. *Keroëides gracilis*, Whitelegge. Part of twig enlarged (× 10). Note the large and regularly arranged spicules.

„ 3. *Keroëides gracilis*, Whitelegge. Portion of main stem enlarged (× 10) showing the smaller and more indefinitely arranged spicules.

„ 4. *Parisis indica*, n. sp. Branch nat. size.

„ 5. *Parisis indica*, n. sp. Part of axis enlarged showing the clusters of epizoic *Palythoa*.

„ 6. *Sympodium tenue*, n. sp. Colony (× 1½) showing the support of siliceous sponge spicules.

„ 7. *Distichoptilum gracile*, Verrill. Part enlarged (× 10). Note the small, scarcely projecting polyps.

„ 8. *Parisis indica*, n. sp. Basal portion (× 1½) showing the disc of attachment.

„ 9. *Parisis indica*, n. sp. Part of a branch enlarged (× 1½). Contrast the proportions of node and internode with those in fig. 8.

„ 10. *Chironephthya macrospiculata*, n. sp. Terminal portion enlarged (× 12). Note the enormous spindles of the cœnenchyma.

„ 11. *Suberogorgia köllikeri*, var. *ceylonensis*, Thomson. Branch nat. size.

„ 12. *Suberogorgia köllikeri*, var. *ceylonensis*, Thomson. Portion enlarged (× 10).

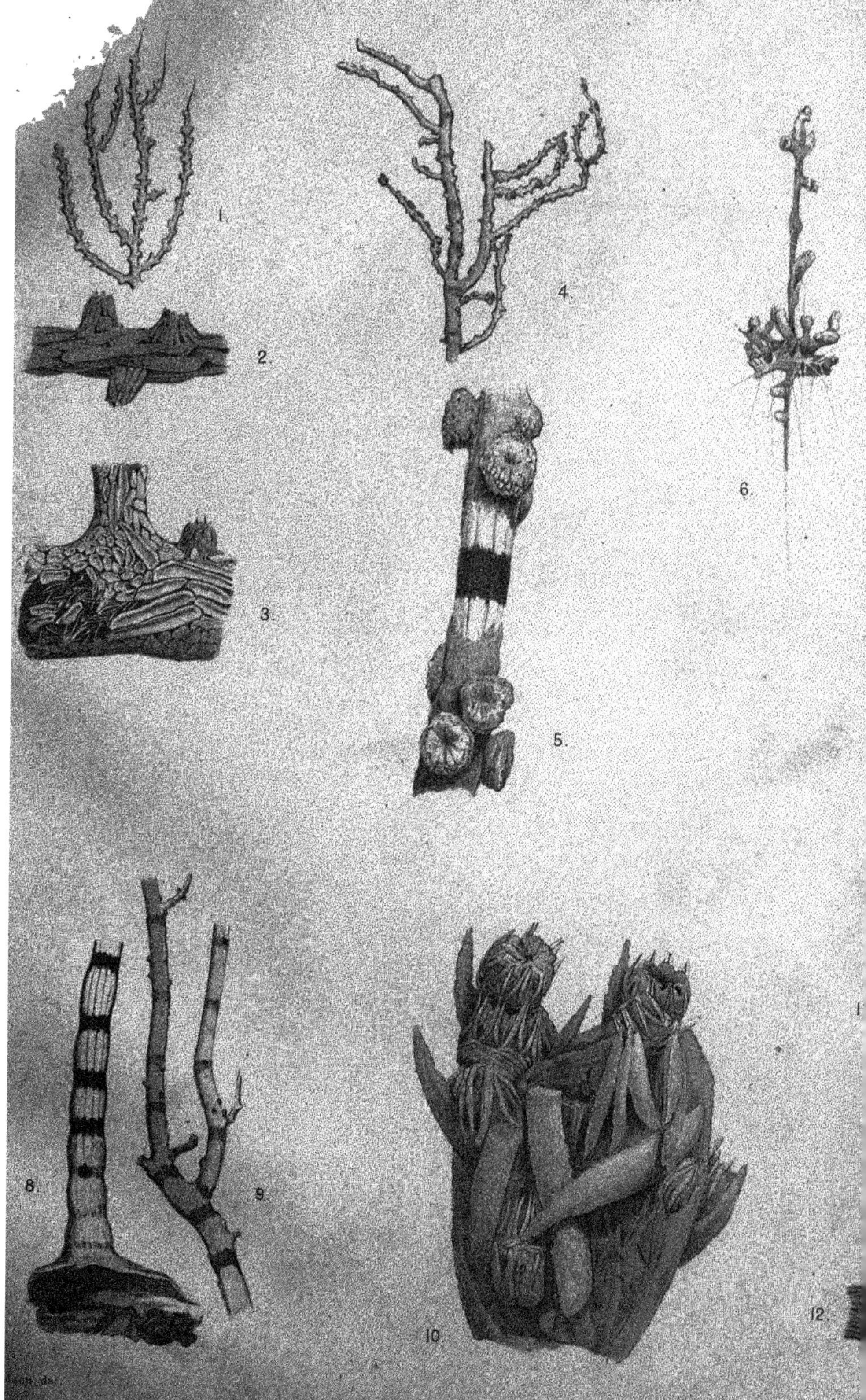
1.
2.
3.
4.
5.
6.
8.
9.
10.
12.

Plate V.

Fig. 1. *Acanthomuricea ramosa*, n. g. et sp. Terminal twig enlarged (× 10). Side to which the polyps are directed.

,, 2. *Stereacanthia indica*, n. g. et sp. Part enlarged (× 10).

,, 3. *Acamptogorgia circium*, n. sp. Twig (× 2). Note the spinose appearance.

,, 4. *Acanthomuricea ramosa*, n. g. et sp. Reverse side of fig. 1.

,, 5. *Umbellula rosea*, n. sp. Rachis nat. size.

,, 6. *Pleurocorallium variabile*, n. sp. Portion enlarged (× 12). Note the arenaceous appearance of the cœnenchyma and the stellate form of the polyps.

,, 7. *Juncella miniacea*, n. sp. Colony nat. size. Note gall.

,, 8. *Acanthomuricea ramosa*, n. g. et sp. Branch (× 1½).

,, 9. *Paragorgia splendens*, n. sp. Branch nat. size.

,, 10. *Astrogorgia rubra*, n. sp. Terminal portion of twig (× 10).

,, 11. *Acis spinosa*, n. sp. Part of twig enlarged (× 14). Note the difference in the spiculation of contiguous portions; also the enormous plate-like spicules.

,, 12. *Juncella miniacea*, n. sp. Part enlarged (× 12) showing the characteristic polyps.

,, 13. *Stenella horrida*, n. sp. Terminal twig enlarged (× 12). Note the rugose appearance.

,, 14. *Paragorgia splendens*, n. sp. Cluster of autozooids enlarged; also showing the arrangement of the siphonozooids.

,, 15. *Acanthogorgia aspera*, Pourtales. Side and end view of polyps (× 10).

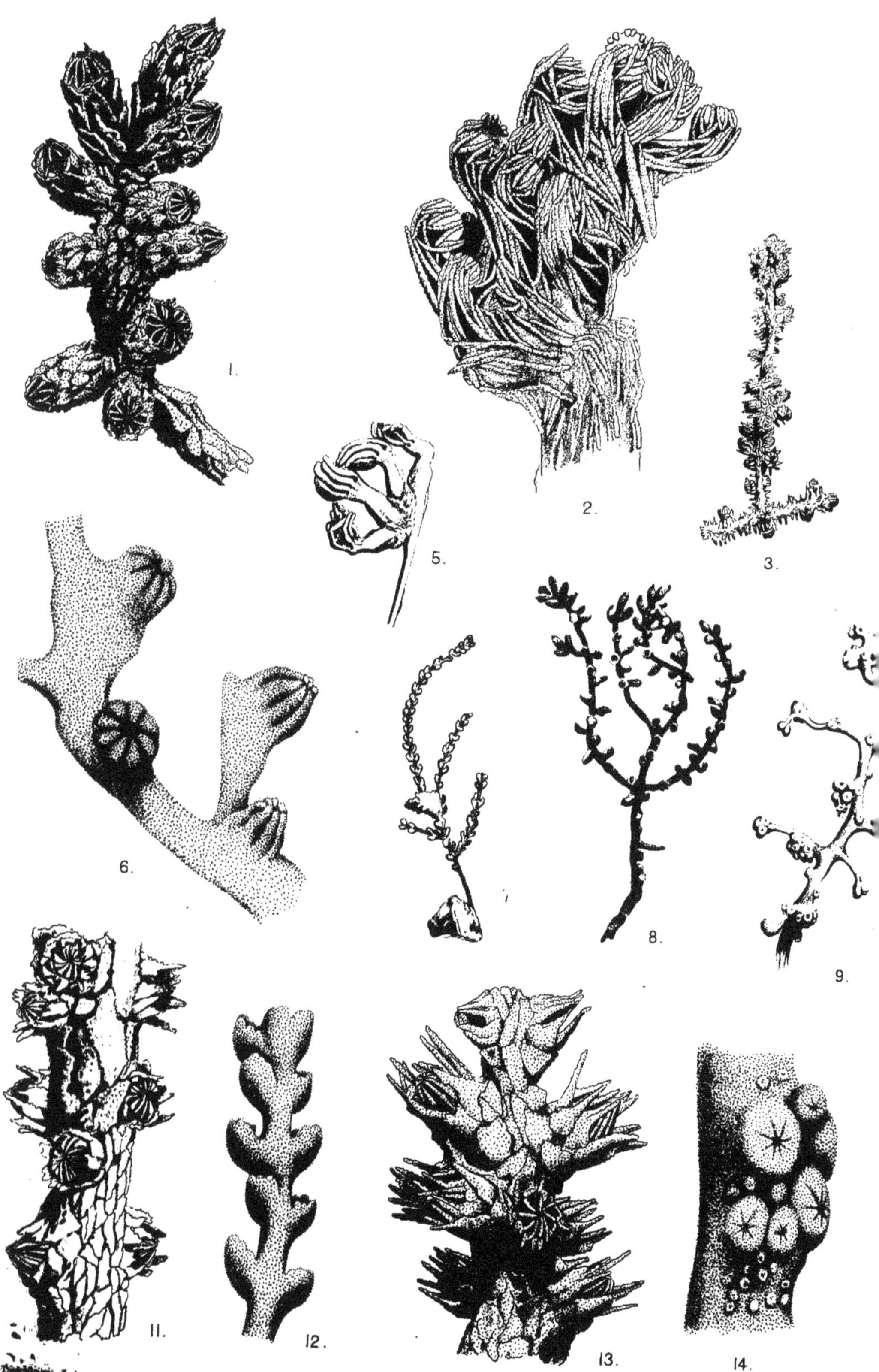
1.
2.
3.
5.
6.
8.
9.
11.
12.
13.
14.

PLATE VII.

FIG. 1. *Pennatula pendula*, n. sp. A pinnule. Note the rugose autozooids and the ova appearing through the semi-transparent walls.

„ 2. *Chrysogorgia orientalis*, Versluys. Portion enlarged to show the mode of branching and the characteristic architecture of the polyps.

„ 3. *Protocaulon indicum*, n. sp. Part enlarged. Note the long polyps and the peculiar flagelliform tentacles.

„ 4. *Bathyptilum indicum*, n. sp. Colony nat. size.

„ 5. *Stachyptilum maculatum*, n. sp. Colony nat. size.

„ 6. *Umbellula elongata*, n. sp. Portion of autozooid cut open to show the embryos.

„ 7. *Protocaulon indicum*, n. sp. Colony nat. size.

„ 8. *Anthoptilum decipiens*, n. sp. Part enlarged showing the reproductive bodies appearing through the thin body-walls.

„ 9. *Stachyptilum maculatum*, n. sp. Part enlarged showing the strong autozooids and the conspicuous stellate siphonozooids.

„ 10. *Funiculina gracilis*, n. sp. Portion enlarged to show the arrangement of the autozooids and prominent projecting eight-lobed siphonozooids in an almost definite row on the metarachidial surface.

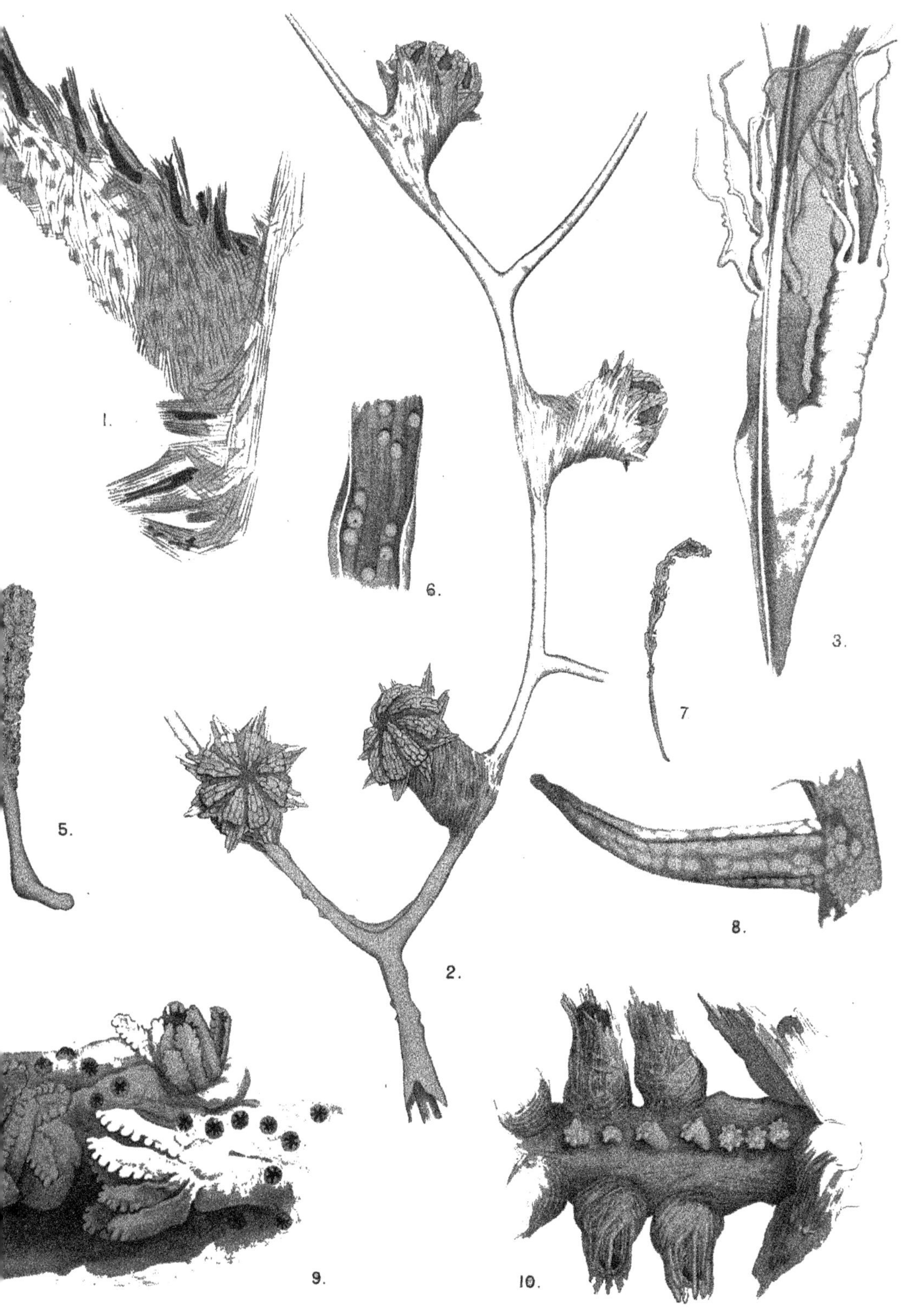

E. Wilson, Cambridge.

PLATE VIII.

FIG. 1. *Pennatula indica*, n. sp. (*a*) Complete colony nat. size; (*b*) portion of rachis enlarged (× 10); (*c*) part of a pinnule enlarged (× 10).

„ 2. *Muricella bengalensis*, n. sp. Twig enlarged (× 8).

„ 3. *Umbellula purpurea*, n. sp. Terminal portion nat. size.

„ 4. *Agaricoides alcocki*, Simpson, n. g. et sp. Colony nat. size.

„ 5. *Pennatula splendens*, n. sp. Portion of rachis with four pinnules (× 4).

„ 6. *Spongodes alcocki*, n. sp. Terminal portion (× 10).

„ 7. *Pennatula pendula*, n. sp. Colony nat. size.

„ 8. *Pennatula veneris*, n. sp. Pinnule (× 6).

„ 9. *Umbellula dura*, n. sp. Colony nat. size.

„ 10. *Pennatula pendula*, n. sp. Pinnule (× 6).

DEEP SEA ALCYONARIA.

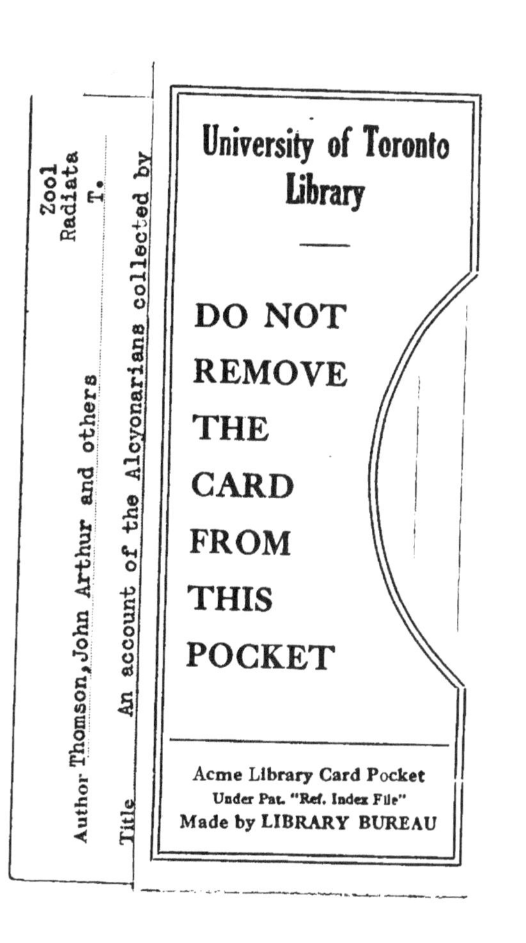

Lightning Source UK Ltd.
Milton Keynes UK
UKHW032253141118
332327UK00005B/206/P